Table of content

Chapter 1: Introduction to TensorFlow

TensorFlow, developed by the Google Brain team, has emerged as a powerful open-source machine learning library that has revolutionized the field of artificial intelligence. Originally released in 2015, TensorFlow has since become a cornerstone for researchers, developers, and businesses seeking to build and deploy machine learning models.

1.1 Definition and Purpose

1.1.1 Definition of TensorFlow

At its core, TensorFlow is an open-source machine learning library designed to facilitate the development and deployment of deep learning models. It is characterized by its flexible architecture, allowing developers to build and train neural networks for a wide range of applications.

1.1.2 Purpose and Goals of TensorFlow

TensorFlow's primary purpose is to simplify the implementation of machine learning algorithms, particularly those involving neural networks. It provides a comprehensive ecosystem that includes tools, libraries, and community resources to support the entire machine learning lifecycle. Whether you're a researcher exploring novel algorithms, a developer building applications, or a business deploying models in production, TensorFlow offers a versatile platform.

1.2 Key Concepts in TensorFlow

1.2.1 Tensors and Operations

TensorFlow revolves around the concept of tensors, which are multi-dimensional arrays representing data. Operations, or mathematical functions, are applied to these tensors to perform computations. Understanding these fundamental concepts is crucial for effectively working with TensorFlow and building complex machine learning models.

1.2.2 Graphs and Sessions

TensorFlow uses a computational graph to represent the flow of data through a model. This graph defines the relationships between tensors and operations. Sessions are used to execute operations within the graph, allowing for efficient computation and optimization.

1.3 TensorFlow Ecosystem

1.3.1 TensorFlow Lite for Mobile and Edge Devices

TensorFlow Lite is an extension designed for mobile and edge devices, enabling the deployment of machine learning models on resource-constrained platforms.

1.3.2 TensorFlow.js for Browser-Based Applications

With TensorFlow.js, developers can bring machine learning capabilities to web browsers, opening up new possibilities for interactive and real-time applications.

1.3.3 TensorFlow Extended (TFX) for Production Pipelines

For businesses deploying machine learning at scale, TensorFlow Extended provides a production-ready platform for managing and orchestrating machine learning workflows.

In this introductory chapter, we will delve into the fundamental concepts of TensorFlow, exploring its architecture, installation, and the broader ecosystem. Whether you are a beginner or an experienced practitioner, this journey through the basics will set the foundation for your exploration of TensorFlow's capabilities.

Chapter 2: Getting Started with TensorFlow

Embarking on the journey with TensorFlow is an exciting step towards mastering the art and science of machine learning. Whether you're a curious beginner or an experienced developer looking to add deep learning skills to your repertoire, this chapter serves as your guide to the essential steps of working with TensorFlow.

2.1 Installing TensorFlow

Before diving into the world of machine learning models and algorithms, it's crucial to set up your development environment. In this section, we will walk you through the process of installing TensorFlow using the pip package manager. From configuring the base installation to verifying its successful setup, you'll gain the necessary foundation to begin your TensorFlow journey.

To install TensorFlow in Python, you can use the **pip** package manager. Here's a simple example of how to install TensorFlow:

Use the following command to install the latest version of TensorFlow

If you have a specific version in mind, replace 'latest' with the desired version number (e.g., '2.7.0')

!pip install tensorflow

2.2 Basics of TensorFlow Syntax

Understanding the syntax and structure of TensorFlow is fundamental to building and training models effectively. This section introduces you to the basic building blocks, such as defining tensors, performing operations, and creating simple computational graphs. By the end of this section, you'll feel confident navigating and writing TensorFlow code.

To illustrate the basics of TensorFlow syntax, let's create a simple example that involves defining tensors, performing operations, and using a session. Note that TensorFlow 2.x typically uses eager execution by default, which means you can execute operations immediately. However, we'll also include an example using a session to highlight the traditional syntax.

import tensorflow as tf

Example 1: Defining and Performing Operations (Eager Execution)

Define two constant tensors

tensor_a = tf.constant([1, 2, 3])

tensor_b = tf.constant([4, 5, 6])

Perform a basic operation (element-wise multiplication)

result = tensor_a * tensor_b

Print the result

```
print("Result using eager execution:", result.numpy())


# Example 2: Using a Session (Traditional Syntax)


# Create a session

with tf.compat.v1.Session() as session:

    # Define two constant tensors

    tensor_a = tf.constant([1, 2, 3])

    tensor_b = tf.constant([4, 5, 6])


    # Perform a basic operation (element-wise multiplication)

    result = session.run(tensor_a * tensor_b)


    # Print the result

    print("Result using a session:", result)
```

In the first example, using eager execution, we define two constant tensors and perform an element-wise multiplication. The result is then printed. In TensorFlow 2.x, eager execution is enabled by default, allowing you to directly evaluate operations.

In the second example, we use a session to perform the same operation. Note that eager execution is not enabled in this case. The **session.run()** method is used to execute the operation and obtain the result.

Feel free to run this code in a Python script or Jupyter notebook to observe the output. TensorFlow's syntax can vary based on the version, and the examples provided here are compatible with TensorFlow 2.x.

2.3 Creating Your First TensorFlow Model

With the basics in place, it's time to take your first steps into model creation. This section guides you through the process of building a simple machine learning model using TensorFlow. From defining the model architecture to compiling and training, you'll witness the power of TensorFlow in action.

In the upcoming sections, we will delve into more advanced topics, including data preprocessing, building neural networks, and optimizing model performance. But before we venture into these realms, mastering the foundational aspects covered in this chapter is crucial.

Whether you're a student, researcher, or developer, this chapter is designed to make your initiation into TensorFlow seamless and enjoyable. By the end of it, you'll have the confidence to explore more intricate machine learning concepts and start crafting your intelligent applications with TensorFlow.

Creating your first TensorFlow model typically involves defining a neural network, compiling it, and training it on a dataset. Below is a simple example of a feedforward neural network for a classification task using TensorFlow's Keras API:

```
import tensorflow as tf

from tensorflow.keras import layers, models


# Example dataset - MNIST digits

(x_train,      y_train),      (x_test,      y_test)      =
tf.keras.datasets.mnist.load_data()

x_train, x_test = x_train / 255.0, x_test / 255.0  # Normalize
pixel values to the range [0, 1]


# Define the model

model = models.Sequential([
```

```
    layers.Flatten(input_shape=(28, 28)),         # Flatten the
input data

    layers.Dense(128, activation='relu'),         # Dense layer
with 128 neurons and ReLU activation

    layers.Dropout(0.2),                          # Dropout layer to
prevent overfitting

    layers.Dense(10, activation='softmax')        # Output layer
with softmax activation for 10 classes

])


# Compile the model

model.compile(optimizer='adam',

              loss='sparse_categorical_crossentropy',

              metrics=['accuracy'])


# Train the model

model.fit(x_train, y_train, epochs=5)


# Evaluate the model on the test set

test_loss, test_accuracy = model.evaluate(x_test, y_test)

print("\nTest Accuracy:", test_accuracy)
```

In this example:

We load the MNIST dataset, which consists of 28x28 pixel grayscale images of handwritten digits (0 to 9).

The model is defined as a sequential stack of layers using the Keras API. It includes a flattening layer, a dense hidden layer with ReLU activation, a dropout layer to prevent overfitting, and an output layer with softmax activation for classification.

The model is compiled with the Adam optimizer, sparse categorical cross-entropy loss (suitable for integer-encoded labels), and accuracy as the metric.

The model is trained on the training dataset for five epochs.

Finally, the model is evaluated on the test set, and the test accuracy is printed.

You can run this code in a Python script or Jupyter notebook to see the training and evaluation process. Adjust the architecture and hyperparameters based on your specific task and dataset.

Chapter 3: Data Preprocessing and Loading with TensorFlow

Data preprocessing lays the groundwork for successful machine learning models. In this chapter, we will explore the essential techniques and best practices for efficiently preparing and loading data using TensorFlow. Whether you are dealing with structured data, images, or text, mastering these skills is key to ensuring your models receive high-quality input.

3.1 Loading and Preparing Data

Before diving into model development, it's crucial to understand how to load and prepare your datasets. This section introduces you to TensorFlow's data loading capabilities, guiding you through the process of loading data from various sources, such as CSV files, databases, and more. Learn how to inspect, clean, and structure your data to create a solid foundation for your machine learning endeavors.

Loading and preparing data in TensorFlow often involves using the **tf.data.Dataset** API. Here's a simple example of

15

loading and preparing data using TensorFlow, specifically the popular MNIST dataset:

```
import tensorflow as tf
```

```
# Load the MNIST dataset
```

```
(x_train,      y_train),      (x_test,      y_test)      =
tf.keras.datasets.mnist.load_data()
```

```
# Normalize pixel values to the range [0, 1]
```

```
x_train, x_test = x_train / 255.0, x_test / 255.0
```

```
# Create a TensorFlow Dataset from the training data
```

```
train_dataset  =  tf.data.Dataset.from_tensor_slices((x_train,
y_train))
```

```
# Shuffle and batch the dataset
```

```
train_dataset                                          =
train_dataset.shuffle(buffer_size=60000).batch(64)
```

```
# Create a TensorFlow Dataset from the test data
```

```
test_dataset  =  tf.data.Dataset.from_tensor_slices((x_test,
y_test))
```

```
# Batch the test dataset
```

```
test_dataset = test_dataset.batch(64)
```

In this example:

We load the MNIST dataset using **tf.keras.datasets.mnist.load_data()**.

Pixel values are normalized to the range [0, 1] to facilitate training.

A **tf.data.Dataset** is created from the training data using **tf.data.Dataset.from_tensor_slices()**.

The training dataset is then shuffled and batched using the **shuffle** and **batch** methods.

A similar process is applied to create and batch the test dataset.

After running this code, you'll have two **tf.data.Dataset** objects (**train_dataset** and **test_dataset**) that you can use for training and evaluation in TensorFlow.

Feel free to adapt this example to your specific dataset and requirements. The **tf.data.Dataset** API provides flexibility for loading and preprocessing data efficiently.

3.2 Handling Missing Data

Missing data can pose challenges to the performance of your models. In this section, we'll explore strategies for identifying and handling missing values in your datasets. From imputation techniques to leveraging TensorFlow functions, you'll gain insights into ensuring the completeness and reliability of your data.

Handling missing data is a crucial step in data preprocessing. While TensorFlow itself does not provide explicit methods for handling missing values, you can use standard Python libraries, such as NumPy or Pandas, in conjunction with TensorFlow. Here's an example using NumPy to handle missing data before creating a TensorFlow dataset:

import numpy as np

import tensorflow as tf

from tensorflow.keras import layers, models

Generate a sample dataset with missing values

data = np.array([[1.0, 2.0, np.nan],

```
        [4.0, np.nan, 6.0],

        [7.0, 8.0, 9.0]])
```

```
# Replace missing values with zeros (or use your preferred imputation method)

data_no_missing = np.nan_to_num(data)
```

```
# Assuming labels are in the last column

features = data_no_missing[:, :-1]

labels = data_no_missing[:, -1]
```

```
# Convert to TensorFlow Dataset

dataset   =   tf.data.Dataset.from_tensor_slices((features, labels))
```

```
# Further processing and model building steps can be added here
```

In this example:

We create a sample dataset (**data**) with missing values represented by **np.nan**.

We use NumPy's **np.nan_to_num** function to replace missing values with zeros. You can choose other imputation strategies depending on your specific requirements.

Features and labels are separated assuming the labels are in the last column.

The dataset is converted into a TensorFlow **Dataset** using **tf.data.Dataset.from_tensor_slices()**.

It's important to note that the handling of missing data often depends on the nature of your dataset and the goals of your machine learning task. You might need more advanced techniques for imputation or handling missing values depending on your specific use case.

3.3 Data Augmentation Techniques

For image-based applications, data augmentation is a powerful tool to enhance model robustness. This section demonstrates how to apply data augmentation using TensorFlow, allowing you to generate diverse training samples from your existing dataset. From rotation to flipping, you'll discover techniques that help your model generalize better to unseen data.

Data augmentation is a common technique to artificially increase the size of your training dataset by applying various transformations to the existing data. TensorFlow provides tools through the **tf.image** module for image-related data augmentation. Here's an example using TensorFlow to perform data augmentation on images:

```
import tensorflow as tf

import matplotlib.pyplot as plt


# Load sample image

image_path = "path_to_your_image.jpg"  # Replace with the path to your image

image = tf.io.read_file(image_path)

image = tf.image.decode_image(image)


# Display original image

plt.figure(figsize=(8, 4))

plt.subplot(1, 2, 1)

plt.imshow(image)

plt.title('Original Image')
```

19

```
# Data augmentation

augmented_image = tf.image.random_flip_left_right(image)

augmented_image                          =
tf.image.random_flip_up_down(augmented_image)

augmented_image                          =
tf.image.random_brightness(augmented_image,
max_delta=0.2)

augmented_image                          =
tf.image.random_contrast(augmented_image,    lower=0.5,
upper=1.5)


# Display augmented image

plt.subplot(1, 2, 2)

plt.imshow(augmented_image)

plt.title('Augmented Image')


plt.show()
```

In this example:

Replace **"path_to_your_image.jpg"** with the actual path to your image.

The original image is loaded and displayed using **tf.io.read_file** and **tf.image.decode_image**.

Data augmentation techniques are applied, including random left-right and up-down flips, random brightness adjustment, and random contrast adjustment.

The original and augmented images are displayed side by side using Matplotlib.

Feel free to experiment with other augmentation functions provided by **tf.image** based on your specific needs. Additionally, when working with datasets, you can integrate data augmentation into your input pipeline using the **tf.data** API.

3.4 Customizing Data Loading for Neural Networks

Neural networks often require specialized preprocessing steps. In this section, we'll delve into preparing data specifically for neural network architectures. Learn how to format input data, handle labels, and create custom data pipelines to seamlessly integrate your data with TensorFlow models.

By the end of this chapter, you'll have a solid understanding of the crucial steps involved in data preprocessing and loading within the TensorFlow framework. These skills will serve as a strong foundation for building robust and effective machine learning models in subsequent chapters.

Let's dive into the world of data preprocessing with TensorFlow and ensure your models are fueled with high-quality, well-organized data!

Customizing data loading for neural networks in TensorFlow often involves creating a custom data pipeline using the **tf.data.Dataset** API. Here's an example of how to customize data loading for a neural network by creating a custom dataset and data augmentation:

```
import tensorflow as tf

import matplotlib.pyplot as plt


# Load sample dataset (e.g., CIFAR-10)

(train_images, train_labels), (test_images, test_labels) = tf.keras.datasets.cifar10.load_data()


# Normalize pixel values to the range [0, 1]

train_images, test_images = train_images / 255.0, test_images / 255.0
```

```python
# Create a custom dataset
train_dataset = tf.data.Dataset.from_tensor_slices((train_images, train_labels))


# Function for data augmentation
def augment_data(image, label):
    # Apply data augmentation techniques
    image = tf.image.random_flip_left_right(image)

    image = tf.image.random_flip_up_down(image)

    image = tf.image.random_brightness(image, max_delta=0.2)

    image = tf.image.random_contrast(image, lower=0.5, upper=1.5)


    return image, label


# Apply data augmentation to the training dataset
train_dataset = train_dataset.map(augment_data)


# Batch and shuffle the training dataset
train_dataset = train_dataset.shuffle(buffer_size=50000).batch(64)


# Create a custom dataset for testing
test_dataset = tf.data.Dataset.from_tensor_slices((test_images, test_labels))

test_dataset = test_dataset.batch(64)
```

```
# Display a batch of augmented images

for images, labels in train_dataset.take(1):

    plt.figure(figsize=(10, 10))

    for i in range(9):

        plt.subplot(3, 3, i + 1)

        plt.imshow(images[i])

        plt.title(f"Class: {labels[i].numpy()}")

        plt.axis("off")

    plt.show()
```

In this example:

We load a sample dataset (CIFAR-10) and normalize pixel values to the range [0, 1].

A custom dataset is created using **tf.data.Dataset.from_tensor_slices**.

The **augment_data** function applies various data augmentation techniques using the **tf.image** module.

Data augmentation is applied to the training dataset using the **map** function.

The training dataset is then batched and shuffled.

A separate dataset is created for testing without data augmentation.

This example uses CIFAR-10 as a sample dataset, and you can adapt the code based on your specific dataset and requirements. Customizing data loading is essential for adapting your neural network to various data formats and preprocessing steps.

Chapter 4: Building Neural Networks with TensorFlow

In the realm of machine learning, neural networks stand as powerful tools for solving complex problems. In this chapter, we will explore the intricacies of building neural networks using TensorFlow, from constructing simple feedforward networks to implementing sophisticated convolutional and recurrent architectures. Whether you're a beginner seeking a foundational understanding or an experienced practitioner aiming to expand your skill set, this chapter serves as your gateway to the world of neural networks with TensorFlow.

4.1 Introduction to Neural Networks

Before we delve into the practicalities, let's establish a solid understanding of neural networks. This section provides an overview of neural network concepts, architectures, and their role in machine learning. Gain insights into the fundamentals of neurons, layers, and the overall structure that defines the power of neural networks.

4.2 Building a Simple Feedforward Neural Network

We'll kick off our exploration by constructing a basic feedforward neural network. Learn how to define the architecture, initialize weights and biases, and implement forward and backward propagation. By the end of this section, you'll have a functional understanding of creating and training a neural network for various tasks.

Certainly! Here's an example of building a simple feedforward neural network using TensorFlow's Keras API:

```
import tensorflow as tf

from tensorflow.keras import layers, models


# Load sample dataset (e.g., MNIST)

(x_train,        y_train),        (x_test,        y_test)        =
tf.keras.datasets.mnist.load_data()


# Normalize pixel values to the range [0, 1]

x_train, x_test = x_train / 255.0, x_test / 255.0


# Create a simple feedforward neural network

model = models.Sequential([

    layers.Flatten(input_shape=(28, 28)),    # Flatten the input
data

    layers.Dense(128, activation='relu'),    # Dense layer with
128 neurons and ReLU activation

    layers.Dropout(0.2),                # Dropout layer to prevent
overfitting

    layers.Dense(10, activation='softmax')  # Output layer with
softmax activation for 10 classes

])
```

```
# Compile the model
model.compile(optimizer='adam',
          loss='sparse_categorical_crossentropy',
          metrics=['accuracy'])


# Train the model
model.fit(x_train, y_train, epochs=5)


# Evaluate the model on the test set
test_loss, test_accuracy = model.evaluate(x_test, y_test)
print("\nTest Accuracy:", test_accuracy)
```

In this example:

We load a sample dataset (MNIST) and normalize pixel values to the range [0, 1].

The model is defined as a sequential stack of layers using the Keras API. It includes a flattening layer, a dense hidden layer with ReLU activation, a dropout layer to prevent overfitting, and an output layer with softmax activation for classification.

The model is compiled with the Adam optimizer, sparse categorical cross-entropy loss (suitable for integer-encoded labels), and accuracy as the metric.

The model is trained on the training dataset for five epochs.

Finally, the model is evaluated on the test set, and the test accuracy is printed.

You can run this code in a Python script or Jupyter notebook to observe the training and evaluation process. Adjust the architecture and hyperparameters based on your specific task and dataset.

4.3 Implementing Convolutional Neural Networks (CNNs)

For tasks involving image analysis, convolutional neural networks (CNNs) are indispensable. In this section, we'll delve into the design and implementation of CNNs using TensorFlow. Explore convolutional layers, pooling, and techniques for extracting meaningful features from images, all while gaining a practical understanding of TensorFlow's capabilities for image processing.

Certainly! Here's an example of implementing a simple Convolutional Neural Network (CNN) using TensorFlow's Keras API. This example uses the CIFAR-10 dataset, which is a common dataset for image classification:

import tensorflow as tf

from tensorflow.keras import layers, models

Load CIFAR-10 dataset

(x_train, y_train), (x_test, y_test) = tf.keras.datasets.cifar10.load_data()

Normalize pixel values to the range [0, 1]

x_train, x_test = x_train / 255.0, x_test / 255.0

Create a simple CNN

```
model = models.Sequential([

    layers.Conv2D(32,       (3,      3),       activation='relu',
input_shape=(32, 32, 3)),  # Convolutional layer with 32 filters
and ReLU activation

    layers.MaxPooling2D((2, 2)),  # Max pooling layer

    layers.Conv2D(64,     (3,    3),    activation='relu'),       #
Convolutional layer with 64 filters and ReLU activation

    layers.MaxPooling2D((2, 2)),  # Max pooling layer

    layers.Conv2D(64,     (3,    3),    activation='relu'),       #
Convolutional layer with 64 filters and ReLU activation

    layers.Flatten(),  # Flatten the output for dense layers

    layers.Dense(64, activation='relu'),  # Dense layer with 64
neurons and ReLU activation

    layers.Dense(10, activation='softmax')  # Output layer with
softmax activation for 10 classes

])


# Compile the model

model.compile(optimizer='adam',

          loss='sparse_categorical_crossentropy',

          metrics=['accuracy'])


# Train the model

model.fit(x_train, y_train, epochs=10)


# Evaluate the model on the test set

test_loss, test_accuracy = model.evaluate(x_test, y_test)

print("\nTest Accuracy:", test_accuracy)
```

In this example:

We load the CIFAR-10 dataset and normalize pixel values to the range [0, 1].

The CNN model is defined using the Keras Sequential API, consisting of convolutional layers, max-pooling layers, and dense layers.

The model is compiled with the Adam optimizer, sparse categorical cross-entropy loss, and accuracy as the metric.

The model is trained on the training dataset for ten epochs.

Finally, the model is evaluated on the test set, and the test accuracy is printed.

You can run this code in a Python script or Jupyter notebook to observe the training and evaluation process. Adjust the architecture and hyperparameters based on your specific task and dataset.

4.4 Recurrent Neural Networks (RNNs)

Sequential data, such as time-series or natural language, requires specialized architectures like recurrent neural networks (RNNs). Learn how to build RNNs using TensorFlow, understanding the nuances of sequential data processing, and unlocking the potential for tasks like language modeling and sentiment analysis.

By the end of this chapter, you'll have the knowledge and confidence to build and experiment with various neural network architectures using TensorFlow. Whether your interests lie in image recognition, natural language processing, or other domains, these foundational skills will propel you forward in your machine learning journey.

Let's dive into the fascinating world of neural networks with TensorFlow and unleash the potential of these powerful models!

Certainly! Here's an example of implementing a simple Recurrent Neural Network (RNN) using TensorFlow's Keras

API. This example uses the IMDB dataset for sentiment analysis:

```
import tensorflow as tf

from tensorflow.keras import layers, models

from tensorflow.keras.datasets import imdb

from tensorflow.keras.preprocessing import sequence


# Load IMDB dataset

max_features = 10000  # Consider the top 10,000 words

max_len = 500  # Limit the sequence length to 500 words

(x_train,      y_train),      (x_test,      y_test)      =
imdb.load_data(num_words=max_features)


# Pad sequences to ensure consistent length

x_train        =        sequence.pad_sequences(x_train,
maxlen=max_len)

x_test = sequence.pad_sequences(x_test, maxlen=max_len)


# Create a simple RNN

model = models.Sequential([

    layers.Embedding(max_features, 32),  # Word embedding
layer

    layers.SimpleRNN(32),  # Simple RNN layer with 32 units

    layers.Dense(1, activation='sigmoid')  # Output layer with
sigmoid activation for binary classification

])


# Compile the model

model.compile(optimizer='adam',

              loss='binary_crossentropy',
```

```
metrics=['accuracy'])


# Train the model

model.fit(x_train,   y_train,   epochs=5,   batch_size=64,
validation_split=0.2)


# Evaluate the model on the test set

test_loss, test_accuracy = model.evaluate(x_test, y_test)

print("\nTest Accuracy:", test_accuracy)
```

In this example:

We load the IMDB dataset and limit the vocabulary to the top 10,000 words.

Sequences are padded to ensure consistent length using the **sequence.pad_sequences** function.

The RNN model is defined using the Keras Sequential API, consisting of an embedding layer, a Simple RNN layer, and an output layer with sigmoid activation for binary classification.

The model is compiled with the Adam optimizer, binary cross-entropy loss, and accuracy as the metric.

The model is trained on the training dataset for five epochs with a batch size of 64 and a validation split of 20%.

Finally, the model is evaluated on the test set, and the test accuracy is printed.

You can run this code in a Python script or Jupyter notebook to observe the training and evaluation process. Adjust the architecture and hyperparameters based on your specific task and dataset.

Chapter 5: Model Training and Evaluation with TensorFlow

Now that we've built our neural network, the next steps involve training the model on our training data and evaluating its performance on unseen data. In this chapter, we'll explore the process of training and evaluating models using TensorFlow, covering essential concepts and practices.

5.1 Training a Neural Network

5.1.1 Defining Training Parameters

Before diving into training, it's crucial to set parameters such as the optimizer, loss function, and evaluation metrics. The choice of these parameters depends on the nature of your task (classification, regression, etc.) and the characteristics of your dataset.

```
model.compile(optimizer='adam',

        loss='sparse_categorical_crossentropy',

        metrics=['accuracy'])
```

5.1.2 Feeding Data to the Model

In TensorFlow, we feed data to the model using the **fit** method. This method takes the training data, labels, batch size, and the number of epochs as arguments.

```
model.fit(x_train, y_train, epochs=10, batch_size=32, validation_data=(x_val, y_val))
```

5.1.3 Monitoring Training Progress

During training, it's essential to monitor the model's performance on both the training and validation sets. TensorFlow allows you to visualize training metrics using tools like TensorBoard.

```
tensorboard_callback = tf.keras.callbacks.TensorBoard(log_dir='./logs')

model.fit(x_train, y_train, epochs=10, batch_size=32, validation_data=(x_val, y_val), callbacks=[tensorboard_callback])
```

5.2 Evaluating a Trained Model

5.2.1 Using Test Data

After training, it's time to evaluate the model's performance on a separate test dataset that it has never seen before. This step provides an unbiased assessment of the model's generalization ability.

```
test_loss, test_accuracy = model.evaluate(x_test, y_test)

print("Test Loss:", test_loss)

print("Test Accuracy:", test_accuracy)
```

5.2.2 Making Predictions

Once the model is trained and evaluated, we can use it to make predictions on new, unseen data.

```
predictions = model.predict(x_new_data)
```

5.2.3 Model Fine-Tuning

Based on evaluation results, you may need to fine-tune your model. This could involve adjusting hyperparameters, modifying the architecture, or collecting more data.

5.3 Overfitting and Regularization

Overfitting is a common challenge during training. Techniques such as dropout layers, L1/L2 regularization, and early stopping can be employed to mitigate overfitting and improve model generalization.

5.4 Summary

In this chapter, we've covered the fundamental steps involved in training and evaluating models using TensorFlow. From setting training parameters to monitoring progress and addressing overfitting, these practices form the backbone of effective model development. As you progress in your machine learning journey, experimenting with different architectures and tuning hyperparameters will become second nature, allowing you to build powerful models that excel in real-world scenarios.

Chapter 6: Advanced TensorFLow Concepts

As you delve deeper into the world of TensorFlow, several advanced concepts and techniques come into play, enhancing your ability to design, train, and deploy sophisticated machine learning models. In this chapter, we'll explore some of these advanced TensorFlow concepts.

6.1 Custom Layers and Models

TensorFlow allows you to create custom layers and models, offering flexibility in designing architectures tailored to specific requirements. This enables the incorporation of custom functionalities or complex neural network components.

```python
class CustomLayer(tf.keras.layers.Layer):

    def __init__(self, units=32):

        super(CustomLayer, self).__init__()

        self.units = units


    def build(self, input_shape):

        self.w = self.add_weight(shape=(input_shape[-1], self.units),

                        initializer='random_normal',

                        trainable=True)

        self.b = self.add_weight(shape=(self.units,),

                        initializer='random_normal',

                        trainable=True)


    def call(self, inputs):
```

35

```
return tf.matmul(inputs, self.w) + self.b
```

```python
# Create a custom model
class CustomModel(tf.keras.Model):
    def __init__(self, units=32):
        super(CustomModel, self).__init__()
        self.custom_layer = CustomLayer(units)
        self.flatten = tf.keras.layers.Flatten()
        self.dense = tf.keras.layers.Dense(10, activation='softmax')

    def call(self, inputs):
        x = self.custom_layer(inputs)
        x = self.flatten(x)
        return self.dense(x)


model = CustomModel()
```

6.2 Transfer Learning

Transfer learning is a technique where a pre-trained model is used as a starting point for a new task. TensorFlow provides access to popular pre-trained models through the Keras Applications module.

```python
base_model = tf.keras.applications.MobileNetV2(input_shape=(224, 224, 3),
include_top=False, weights='imagenet')


# Freeze the base model
base_model.trainable = False


# Create a new model on top
model = tf.keras.Sequential([
    base_model,
```

```
tf.keras.layers.GlobalAveragePooling2D(),

tf.keras.layers.Dense(10, activation='softmax')

])
```

6.3 TensorFlow Serving

TensorFlow Serving is a flexible, high-performance serving system for machine learning models designed for production environments. It allows you to deploy models in a server and serve predictions through a simple API.

```
# Example serving command

docker run -p 8501:8501 --name=tf_model_serving --mount
type=bind,source=/path/to/saved_model,target=/models/model -e
MODEL_NAME=model -t tensorflow/serving
```

6.4 TensorFlow Lite

TensorFlow Lite is a lightweight solution for deploying machine learning models on mobile and edge devices. It optimizes models for resource-constrained environments without compromising performance.

```
# Convert a model to TensorFlow Lite format

converter = tf.lite.TFLiteConverter.from_keras_model(model)

tflite_model = converter.convert()


# Save the TensorFlow Lite model to a file

with open('model.tflite', 'wb') as f:

    f.write(tflite_model)
```

6.5 Distributed Training

TensorFlow supports distributed training across multiple devices or machines, enabling the training of large models on large datasets efficiently.

```
# Example of distributed training using tf.distribute.MirroredStrategy

strategy = tf.distribute.MirroredStrategy()

with strategy.scope():

    model = create_your_model()

    model.compile(optimizer='adam', loss='sparse_categorical_crossentropy',
metrics=['accuracy'])


# Continue with the usual model.fit(...) for training
```

6.6 TensorFlow Extended (TFX)

TensorFlow Extended is an end-to-end platform for deploying production-ready machine learning models. It includes components for data validation, model analysis, and continuous evaluation.

```
# Example TFX pipeline

# (Note: TFX pipeline setup is beyond the scope of a code snippet)

import tfx


pipeline = tfx.dsl.Pipeline(

    ...  # Define your TFX components and pipeline steps here

)


# Run the TFX pipeline

tfx.orchestration.LocalDagRunner().run(pipeline)
```

6.7 Summary

This chapter has introduced several advanced TensorFlow concepts, showcasing the versatility and power of the TensorFlow ecosystem. Custom layers, transfer learning, deployment with TensorFlow Serving, TensorFlow Lite for edge devices, distributed training, and TensorFlow Extended for end-to-end machine learning pipelines are among the key components that can take your machine learning projects to the next level. As you continue your journey in machine learning, these advanced concepts will play a crucial role in tackling complex challenges and building robust, production-ready solutions.

Chapter 7: TensorFlow for Natural Language Processing (NLP)

Natural Language Processing (NLP) is a field of artificial intelligence that focuses on the interaction between computers and humans through natural language. TensorFlow, with its rich ecosystem and powerful capabilities, plays a pivotal role in advancing NLP research and applications. In this chapter, we'll explore how TensorFlow can be leveraged for various NLP tasks.

7.1 Tokenization and Word Embeddings

Tokenization is a crucial step in NLP, breaking down text into individual units such as words or subwords. TensorFlow provides tools for efficient tokenization and embedding generation.

```
import tensorflow as tf

from tensorflow.keras.preprocessing.text import Tokenizer

from tensorflow.keras.preprocessing.sequence import pad_sequences


# Tokenization example

texts = ["This is a positive sentence.", "This is a negative sentence."]

tokenizer = Tokenizer()

tokenizer.fit_on_texts(texts)

sequences = tokenizer.texts_to_sequences(texts)
```

```
# Padding sequences for uniform length

padded_sequences = pad_sequences(sequences)
```

7.1.1 Project for Tokenization and Word Embeddings

In this Python project, we'll focus on tokenization and word embeddings using TensorFlow. We'll use the IMDb movie reviews dataset and create a simple model that tokenizes text, converts it to word embeddings, and performs sentiment analysis.

Project: Tokenization and Word Embeddings with TensorFlow

Step 1: Install TensorFlow

pip install tensorflow

Step 2: Import Libraries

import tensorflow as tf

from tensorflow.keras.preprocessing.text import Tokenizer

from tensorflow.keras.preprocessing.sequence import pad_sequences

from tensorflow.keras.models import Sequential

from tensorflow.keras.layers import Embedding, Flatten, Dense

Step 3: Load and Preprocess IMDb Dataset

```
# Load the IMDb dataset
```

imdb = tf.keras.datasets.imdb

(train_data, train_labels), (test_data, test_labels) = imdb.load_data(num_words=10000)


```
# Preprocess the data
```

max_len = 100

```
train_data = pad_sequences(train_data, maxlen=max_len)
```

```
test_data = pad_sequences(test_data, maxlen=max_len)
```

Step 4: Tokenization

```
# Tokenization
```

```
tokenizer           =           Tokenizer(num_words=10000,
oov_token='<OOV>')
```

```
tokenizer.fit_on_texts(train_data)
```

```
# Convert text to sequences
```

```
train_sequences = tokenizer.texts_to_sequences(train_data)
```

```
test_sequences = tokenizer.texts_to_sequences(test_data)
```

Step 5: Word Embeddings Model

```
# Word Embeddings Model
```

```
embedding_dim = 16
```

```
model = Sequential()
```

```
model.add(Embedding(10000,           embedding_dim,
input_length=max_len))
```

```
model.add(Flatten())
```

```
model.add(Dense(1, activation='sigmoid'))
```

Step 6: Compile and Train the Model

```
# Compile the model
```

```
model.compile(optimizer='adam',
```

```
        loss='binary_crossentropy',
```

```
        metrics=['accuracy'])
```

```
# Train the model
```

```
model.fit(train_sequences,           train_labels,           epochs=5,
validation_data=(test_sequences, test_labels))
```

Step 7: Evaluate the Model

Evaluate the model

test_loss, test_acc = model.evaluate(test_sequences, test_labels, verbose=2)

print(f'Test accuracy: {test_acc}')

How to Run the Project:

Save the above code in a Python file (e.g., **word_embeddings.py**).

Open a terminal and navigate to the directory containing the file.

Run the script using **python word_embeddings.py**.

This project demonstrates tokenization and word embeddings using TensorFlow with a simple sentiment analysis model. You can further experiment by adjusting parameters, exploring different datasets, or incorporating more advanced NLP techniques.

7.2 Word Embeddings with TensorFlow Hub

Word embeddings capture semantic relationships between words. TensorFlow Hub offers pre-trained embeddings that can be easily integrated into your NLP models.

import tensorflow as tf

import tensorflow_hub as hub

Load pre-trained embedding model from TensorFlow Hub

embedding_layer = hub.KerasLayer("https://tfhub.dev/google/tf2-preview/gnews-swivel-20dim/1", output_shape=[20], input_shape=[], dtype=tf.string)

7.2.1 Project: Word Embeddings with TensorFlow Hub

In this Python project, we'll use TensorFlow Hub to leverage pre-trained word embeddings for a text classification task. We'll focus on sentiment analysis using the IMDb movie reviews dataset.

Step 1: Install TensorFlow and TensorFlow Hub

pip install tensorflow tensorflow-hub

Step 2: Import Libraries

import tensorflow as tf

import tensorflow_hub as hub

from tensorflow.keras.preprocessing.text import Tokenizer

from tensorflow.keras.preprocessing.sequence import pad_sequences

from tensorflow.keras.models import Sequential

from tensorflow.keras.layers import Embedding, Flatten, Dense

Step 3: Load and Preprocess IMDb Dataset

Load the IMDb dataset

imdb = tf.keras.datasets.imdb

(train_data, train_labels), (test_data, test_labels) = imdb.load_data(num_words=10000)

Preprocess the data

max_len = 100

train_data = pad_sequences(train_data, maxlen=max_len)

test_data = pad_sequences(test_data, maxlen=max_len)

Step 4: Tokenization

Tokenization

```
tokenizer                =              Tokenizer(num_words=10000,
oov_token='<OOV>')
```

```
tokenizer.fit_on_texts(train_data)
```

Convert text to sequences

```
train_sequences = tokenizer.texts_to_sequences(train_data)
```

```
test_sequences = tokenizer.texts_to_sequences(test_data)
```

Step 5: Load TensorFlow Hub Word Embeddings

```
# Load pre-trained word embeddings from TensorFlow Hub
```

```
embedding_url        =            "https://tfhub.dev/google/tf2-
preview/gnews-swivel-20dim/1"
```

```
hub_layer = hub.KerasLayer(embedding_url, input_shape=[],
dtype=tf.string, trainable=True)
```

Step 6: Word Embeddings Model

```
# Word Embeddings Model
```

```
model = Sequential()
```

```
model.add(hub_layer)
```

```
model.add(Dense(16, activation='relu'))
```

```
model.add(Dense(1, activation='sigmoid'))
```

Step 7: Compile and Train the Model

```
# Compile the model
```

```
model.compile(optimizer='adam',
```

```
        loss='binary_crossentropy',
```

```
        metrics=['accuracy'])
```

```
# Train the model
```

```
model.fit(train_data,          train_labels,          epochs=5,
validation_data=(test_data, test_labels))
```

Step 8: Evaluate the Model

```
# Evaluate the model

test_loss, test_acc = model.evaluate(test_data, test_labels,
verbose=2)

print(f'Test accuracy: {test_acc}')
```

How to Run the Project:

Save the above code in a Python file (e.g., **word_embeddings_tf_hub.py**).

Open a terminal and navigate to the directory containing the file.

Run the script using **python word_embeddings_tf_hub.py**.

This project demonstrates using TensorFlow Hub to incorporate pre-trained word embeddings for text classification. Feel free to experiment with different pre-trained embeddings from TensorFlow Hub or adjust the model architecture based on your specific NLP task.

7.3 Text Classification

Text classification involves assigning predefined categories or labels to text documents. TensorFlow simplifies the creation of text classification models using high-level APIs.

```
import tensorflow as tf

from tensorflow.keras import layers, models


# Example text classification model

model = models.Sequential([

    layers.Embedding(input_dim=vocab_size,
output_dim=embedding_dim, input_length=max_length),

    layers.GlobalAveragePooling1D(),

    layers.Dense(64, activation='relu'),

    layers.Dense(num_classes, activation='softmax')
```

])

7.3.1 Project: Text Classification with TensorFlow

In this Python project, we'll build a text classification model using TensorFlow. We'll use the IMDb movie reviews dataset for sentiment analysis. The goal is to classify movie reviews as positive or negative.

Step 1: Install TensorFlow

pip install tensorflow

Step 2: Import Libraries

import tensorflow as tf

from tensorflow.keras.datasets import imdb

from tensorflow.keras.preprocessing.text import Tokenizer

from tensorflow.keras.preprocessing.sequence import pad_sequences

from tensorflow.keras.models import Sequential

from tensorflow.keras.layers import Embedding, Flatten, Dense

Step 3: Load and Preprocess IMDb Dataset

Load the IMDb dataset

num_words = 10000

(train_data, train_labels), (test_data, test_labels) = imdb.load_data(num_words=num_words)

Preprocess the data

max_len = 100

train_data = pad_sequences(train_data, maxlen=max_len)

test_data = pad_sequences(test_data, maxlen=max_len)

Step 4: Tokenization

```
# Tokenization

tokenizer = Tokenizer(num_words=num_words,
oov_token='<OOV>')

tokenizer.fit_on_texts(train_data)


# Convert text to sequences

train_sequences = tokenizer.texts_to_sequences(train_data)

test_sequences = tokenizer.texts_to_sequences(test_data)
```

Step 5: Text Classification Model

```
# Text Classification Model

model = Sequential()

model.add(Embedding(num_words, 16,
input_length=max_len))

model.add(Flatten())

model.add(Dense(16, activation='relu'))

model.add(Dense(1, activation='sigmoid'))
```

Step 6: Compile and Train the Model

```
# Compile the model

model.compile(optimizer='adam',

        loss='binary_crossentropy',

        metrics=['accuracy'])


# Train the model

model.fit(train_sequences, train_labels, epochs=5,
validation_data=(test_sequences, test_labels))
```

Step 7: Evaluate the Model

```
# Evaluate the model

test_loss, test_acc = model.evaluate(test_sequences,
test_labels, verbose=2)
```

```
print(f'Test accuracy: {test_acc}')
```

How to Run the Project:

Save the above code in a Python file (e.g., **text_classification.py**).

Open a terminal and navigate to the directory containing the file.

Run the script using **python text_classification.py**.

This project demonstrates a simple text classification task using a neural network with TensorFlow. Feel free to experiment with different architectures, hyperparameters, or even explore advanced techniques such as using pre-trained word embeddings or recurrent neural networks for improved performance.

7.4 Named Entity Recognition (NER)

Named Entity Recognition identifies entities such as names, locations, and organizations in text. TensorFlow allows for the creation of custom NER models.

```
import tensorflow as tf

from tensorflow.keras import layers, models


# Example NER model using LSTM

model = models.Sequential([

    layers.Embedding(input_dim=vocab_size,    output_dim=embedding_dim, input_length=max_length),

    layers.Bidirectional(layers.LSTM(64, return_sequences=True)),

    layers.Dense(num_classes, activation='softmax')

])
```

7.4.1 Project: Named Entity Recognition (NER) with TensorFlow and spaCy

Named Entity Recognition (NER) is a common Natural Language Processing (NLP) task where the goal is to identify and classify named entities (such as persons, organizations, and locations) in text. In this Python project, we'll build a Named Entity Recognition model using TensorFlow and the spaCy library.

Step 1: Install Dependencies

pip install tensorflow spacy

python -m spacy download en_core_web_sm

Step 2: Import Libraries

import tensorflow as tf

import spacy

from tensorflow.keras.models import Sequential

from tensorflow.keras.layers import Embedding, Bidirectional, LSTM, Dense

Step 3: Load and Preprocess NER Dataset

For this example, we'll use a sample dataset. In practice, you'd use a more extensive labeled dataset.

Sample dataset

sentences = [

   "Apple Inc. is planning to open a new store in New York City.",

   "John works at Microsoft Corporation in Seattle."

]

Labels for each word (B: Beginning, I: Inside, O: Outside)

labels = [

   ['B-org', 'I-org', 'O', 'O', 'O', 'O', 'B-geo', 'O', 'O', 'O', 'B-geo', 'I-geo', 'O'],

```
['B-per', 'O', 'O', 'B-org', 'I-org', 'O', 'B-geo', 'O']
]
```

Step 4: Preprocess and Tokenize Text

```
# Load spaCy for tokenization
nlp = spacy.load("en_core_web_sm")


# Tokenize and create a vocabulary
words = set()
for sentence in sentences:
    doc = nlp(sentence)
    words.update(token.text.lower() for token in doc)


# Create word-to-index mapping
word_to_index = {word: idx + 1 for idx, word in enumerate(words)}
index_to_word = {idx + 1: word for idx, word in enumerate(words)}


# Convert sentences and labels to sequences
X = [[word_to_index[word.text.lower()] for word in nlp(sentence)] for sentence in sentences]
y = [[labels_to_index[label] for label in sentence_labels] for sentence_labels in labels]
```

Step 5: Named Entity Recognition Model

```
# Named Entity Recognition Model
model = Sequential()
model.add(Embedding(input_dim=len(words) + 1, output_dim=16, input_length=max_len))
model.add(Bidirectional(LSTM(64, return_sequences=True)))
```

```
model.add(Dense(len(labels_to_index)         +         1,
activation='softmax'))
```

Compile the model

```
model.compile(optimizer='adam',
loss='sparse_categorical_crossentropy', metrics=['accuracy'])
```

Step 6: Train the Model

Train the model

```
model.fit(X, y, epochs=10, batch_size=1)
```

How to Run the Project:

Save the above code in a Python file (e.g., **named_entity_recognition.py**).

Open a terminal and navigate to the directory containing the file.

Run the script using **python named_entity_recognition.py**.

This project provides a basic structure for building a Named Entity Recognition model using TensorFlow and spaCy. You can further enhance the project by using a more extensive dataset, experimenting with different architectures, and fine-tuning hyperparameters for improved performance.

7.5 Sequence-to-Sequence Models

Sequence-to-Sequence (Seq2Seq) models are essential for tasks like machine translation. TensorFlow's Keras API simplifies the implementation of Seq2Seq models.

```
import tensorflow as tf

from tensorflow.keras import layers, models


# Example Seq2Seq model for machine translation
```

```
encoder_inputs = layers.Input(shape=(None,))

encoder_embedding        =        layers.Embedding(input_dim=vocab_size,
output_dim=embedding_dim)(encoder_inputs)

encoder_lstm,       state_h,       state_c       =        layers.LSTM(64,
return_state=True)(encoder_embedding)

encoder_states = [state_h, state_c]


decoder_inputs = layers.Input(shape=(None,))

decoder_embedding        =        layers.Embedding(input_dim=vocab_size,
output_dim=embedding_dim)(decoder_inputs)

decoder_lstm                    =                    layers.LSTM(64,
return_sequences=True)(decoder_embedding,
initial_state=encoder_states)

decoder_outputs          =          layers.Dense(vocab_size,
activation='softmax')(decoder_lstm)


model  =  models.Model([encoder_inputs,  decoder_inputs],
decoder_outputs)
```

7.5.1 Project: Sequence-to-Sequence Model for Sequence Reversal

Sequence-to-Sequence (Seq2Seq) models are commonly used for tasks like machine translation, summarization, and more. In this Python project, we'll build a basic Seq2Seq model using TensorFlow for a simple task: reversing sequences.

Step 1: Install TensorFlow

pip install tensorflow

Step 2: Import Libraries

import numpy as np

import tensorflow as tf

from tensorflow.keras.models import Sequential

from tensorflow.keras.layers import Embedding, LSTM, RepeatVector, Dense

53

Step 3: Generate Reversal Dataset

```python
# Generate sequences for reversal task

def generate_sequences(num_samples, max_length):

    sequences = np.random.randint(1, 10, size=(num_samples, max_length))

    reversed_sequences = np.array([list(reversed(seq)) for seq in sequences])

    return sequences, reversed_sequences


num_samples = 10000

max_length = 10

input_sequences, target_sequences = generate_sequences(num_samples, max_length)
```

Step 4: Preprocess Data

```python
# Preprocess data

vocab_size = 10

embedding_dim = 8


model_input = tf.keras.preprocessing.sequence.pad_sequences(input_sequences, maxlen=max_length, padding='post')

model_target = tf.keras.preprocessing.sequence.pad_sequences(target_sequences, maxlen=max_length, padding='post')
```

Step 5: Sequence-to-Sequence Model

```python
# Sequence-to-Sequence Model

model = Sequential()

model.add(Embedding(vocab_size, embedding_dim, input_length=max_length))

model.add(LSTM(100))

model.add(RepeatVector(max_length))
```

```
model.add(LSTM(100, return_sequences=True))

model.add(Dense(vocab_size, activation='softmax'))


# Compile the model

model.compile(optimizer='adam',
loss='categorical_crossentropy', metrics=['accuracy'])
```

Step 6: Train the Model

```
# One-hot encode target sequences

model_target_one_hot                                    =
tf.keras.utils.to_categorical(model_target,
num_classes=vocab_size)


# Train the model

model.fit(model_input,   model_target_one_hot,   epochs=10,
batch_size=64, validation_split=0.2)
```

Step 7: Evaluate the Model

```
# Evaluate the model on new sequences

test_input_sequences,        test_target_sequences      =
generate_sequences(5, max_length)

test_input_sequences_padded                             =
tf.keras.preprocessing.sequence.pad_sequences(test_input_
sequences, maxlen=max_length, padding='post')

test_target_sequences_one_hot                           =
tf.keras.utils.to_categorical(test_target_sequences,
num_classes=vocab_size)


# Evaluate the model

test_loss,              test_accuracy                   =
model.evaluate(test_input_sequences_padded,
test_target_sequences_one_hot, verbose=2)

print(f'Test Accuracy: {test_accuracy * 100:.2f}%')
```

How to Run the Project:

Save the above code in a Python file (e.g., **seq2seq_reversal.py**).

Open a terminal and navigate to the directory containing the file.

Run the script using **python seq2seq_reversal.py**.

This project demonstrates a basic Seq2Seq model for the sequence reversal task. You can modify and extend this project for more complex tasks like machine translation by using appropriate datasets and adjusting the model architecture accordingly.

7.6 Transformers with TensorFlow

Transformers, introduced by the "Attention is All You Need" paper, have become the backbone of many state-of-the-art NLP models. TensorFlow provides tools for building and using transformer-based models.

```python
import tensorflow as tf

from tensorflow.keras import layers, models

from transformers import TFAutoModel, AutoTokenizer


# Example transformer-based model using Hugging Face's transformers library

tokenizer = AutoTokenizer.from_pretrained("bert-base-uncased")

bert_model = TFAutoModel.from_pretrained("bert-base-uncased")


# Build a custom model using the transformer layer

inputs = layers.Input(shape=(max_length,), dtype=tf.int32)

outputs = bert_model(inputs)[0]

outputs = layers.GlobalAveragePooling1D()(outputs)

outputs = layers.Dense(num_classes, activation='softmax')(outputs)
```

```
model = models.Model(inputs, outputs)
```

7.6.1 Project: Text Classification with Transformers

Transformers have become a pivotal architecture in natural language processing and beyond. In this Python project, we'll build a simple text classification model using the Transformer architecture with TensorFlow.

Step 1: Install Dependencies

pip install tensorflow transformers

Step 2: Import Libraries

import tensorflow as tf

from transformers import TFBertModel, BertTokenizer

from tensorflow.keras.models import Sequential

from tensorflow.keras.layers import Dense, GlobalAveragePooling1D

Step 3: Load and Preprocess Data

For this example, we'll use the IMDb movie reviews dataset for sentiment analysis.

Load IMDb dataset

imdb = tf.keras.datasets.imdb

(train_data, train_labels), (test_data, test_labels) = imdb.load_data(num_words=10000)

Preprocess the data

max_len = 128

train_data = tf.keras.preprocessing.sequence.pad_sequences(train_data, maxlen=max_len, padding='post')

```
test_data                                    =
tf.keras.preprocessing.sequence.pad_sequences(test_data,
maxlen=max_len, padding='post')
```

Step 4: Load BERT Model and Tokenizer

```
# Load BERT model and tokenizer

bert_model   =   TFBertModel.from_pretrained('bert-base-
uncased')

tokenizer   =   BertTokenizer.from_pretrained('bert-base-
uncased')
```

Step 5: Tokenize and Convert Text Data

```
# Tokenize and convert text data

def tokenize_data(data):

    return            tokenizer(data,          truncation=True,
padding='max_length',                   max_length=max_len,
return_tensors='tf')


train_data_bert = tokenize_data(train_data)

test_data_bert = tokenize_data(test_data)
```

Step 6: Build and Train the Model

```
# Build and train the model

model = Sequential([

    bert_model,

    GlobalAveragePooling1D(),

    Dense(64, activation='relu'),

    Dense(1, activation='sigmoid')

])


# Compile the model

model.compile(optimizer='adam', loss='binary_crossentropy',
metrics=['accuracy'])
```

Train the model

```
model.fit(train_data_bert,        train_labels,        epochs=3,
validation_data=(test_data_bert, test_labels))
```

Step 7: Evaluate the Model

Evaluate the model

```
test_loss,    test_acc    =    model.evaluate(test_data_bert,
test_labels, verbose=2)
```

```
print(f'Test accuracy: {test_acc}')
```

How to Run the Project:

Save the above code in a Python file (e.g., **transformer_text_classification.py**).

Open a terminal and navigate to the directory containing the file.

Run the script using **python transformer_text_classification.py**.

This project demonstrates using the Transformer architecture (BERT) for text classification. You can customize the model for other NLP tasks, experiment with different pre-trained Transformer models, and fine-tune hyperparameters for improved performance.

7.7 Summary

TensorFlow has become a cornerstone in the field of Natural Language Processing, providing a versatile and efficient framework for developing NLP models. From tokenization and embeddings to complex tasks like text classification, Named Entity Recognition, and sequence-to-sequence modeling, TensorFlow's rich ecosystem empowers researchers and developers to address a wide array of NLP challenges. As you embark on NLP projects, TensorFlow's flexibility and scalability will prove instrumental in building cutting-edge models for understanding and generating human language.

Chapter 8: TensorFlow for Computer Vision

Computer Vision involves teaching machines to interpret and understand visual information from the world. TensorFlow, with its comprehensive set of tools and libraries, has emerged as a powerful platform for developing state-of-the-art Computer Vision models. In this chapter, we'll explore how TensorFlow can be harnessed for various Computer Vision tasks.

8.1 Image Preprocessing and Augmentation

TensorFlow provides utilities for efficient image preprocessing and augmentation, essential for preparing datasets and enhancing model generalization.

```
import tensorflow as tf

from tensorflow.keras.preprocessing.image import ImageDataGenerator


# Example image augmentation

datagen = ImageDataGenerator(

    rotation_range=20,

    width_shift_range=0.2,

    height_shift_range=0.2,
```

```
shear_range=0.2,

zoom_range=0.2,

horizontal_flip=True,

fill_mode='nearest'
```

)

8.1.1 Project: Image Preprocessing and Augmentation with TensorFlow

Image preprocessing and augmentation are essential steps in training deep learning models for image recognition. In this Python project, we'll use TensorFlow and the Keras API to demonstrate image preprocessing and augmentation for a simple image classification task.

Step 1: Install TensorFlow

pip install tensorflow\

Step 2: Import Libraries

import tensorflow as tf

from tensorflow.keras.preprocessing.image import ImageDataGenerator

import matplotlib.pyplot as plt

Step 3: Load and Preprocess Image Data

For this example, we'll use the CIFAR-10 dataset.

Load CIFAR-10 dataset

(train_images, train_labels), (test_images, test_labels) = tf.keras.datasets.cifar10.load_data()

Normalize pixel values to be between 0 and 1

train_images, test_images = train_images / 255.0, test_images / 255.0

Step 4: Visualize Original Images

```
# Visualize original images

plt.figure(figsize=(10, 10))

for i in range(25):

    plt.subplot(5, 5, i+1)

    plt.xticks([])

    plt.yticks([])

    plt.grid(False)

    plt.imshow(train_images[i], cmap=plt.cm.binary)

plt.show()
```

Step 5: Image Augmentation

```
# Image augmentation

datagen = ImageDataGenerator(

    rotation_range=40,

    width_shift_range=0.2,

    height_shift_range=0.2,

    shear_range=0.2,

    zoom_range=0.2,

    horizontal_flip=True,

    fill_mode='nearest'

)


# Visualize augmented images

img = train_images[0]

img = img.reshape((1,) + img.shape)


plt.figure(figsize=(10, 10))

for batch in datagen.flow(img, batch_size=1):
```

```
plt.subplot(5, 5, 1)

plt.xticks([])

plt.yticks([])

plt.grid(False)

plt.imshow(batch[0], cmap=plt.cm.binary)

break
```

plt.show()

How to Run the Project:

Save the above code in a Python file (e.g., **image_preprocessing_augmentation.py**).

Open a terminal and navigate to the directory containing the file.

Run the script using **python image_preprocessing_augmentation.py**.

This project demonstrates image preprocessing and augmentation using TensorFlow. You can adjust the parameters of the **ImageDataGenerator** class for different augmentation techniques and customize it based on your specific dataset and task. Image augmentation is especially useful for increasing the diversity of your training data and improving the generalization performance of your models.

8.2 Convolutional Neural Networks (CNNs)

Convolutional Neural Networks (CNNs) are the backbone of many Computer Vision models. TensorFlow's Keras API simplifies the creation of CNN architectures.

```
import tensorflow as tf

from tensorflow.keras import layers, models
```

```
# Example CNN model

model = models.Sequential([

    layers.Conv2D(32, (3, 3), activation='relu', input_shape=(224, 224, 3)),

    layers.MaxPooling2D((2, 2)),

    layers.Conv2D(64, (3, 3), activation='relu'),

    layers.MaxPooling2D((2, 2)),

    layers.Conv2D(64, (3, 3), activation='relu'),

    layers.Flatten(),

    layers.Dense(64, activation='relu'),

    layers.Dense(10, activation='softmax')

])
```

8.2.1 Project: Convolutional Neural Network (CNN) for Image Classification

In this Python project, we'll build a Convolutional Neural Network (CNN) using TensorFlow for a simple image classification task. We'll use the CIFAR-10 dataset.

Step 1: Install TensorFlow

pip install tensorflow

Step 2: Import Libraries

import tensorflow as tf

from tensorflow.keras import datasets, layers, models

import matplotlib.pyplot as plt

Step 3: Load and Preprocess CIFAR-10 Dataset

Load CIFAR-10 dataset

(train_images, train_labels), (test_images, test_labels) = datasets.cifar10.load_data()

Normalize pixel values to be between 0 and 1

train_images, test_images = train_images / 255.0, test_images / 255.0

Step 4: Visualize Images

```
# Visualize a few images
plt.figure(figsize=(10, 10))
for i in range(25):
    plt.subplot(5, 5, i+1)
    plt.xticks([])
    plt.yticks([])
    plt.grid(False)
    plt.imshow(train_images[i], cmap=plt.cm.binary)
    plt.xlabel(train_labels[i][0])
plt.show()
```

Step 5: Build CNN Model

```
# Build CNN model
model = models.Sequential()
model.add(layers.Conv2D(32,    (3,    3),    activation='relu',
input_shape=(32, 32, 3)))
model.add(layers.MaxPooling2D((2, 2)))
model.add(layers.Conv2D(64, (3, 3), activation='relu'))
model.add(layers.MaxPooling2D((2, 2)))
model.add(layers.Conv2D(64, (3, 3), activation='relu'))


model.add(layers.Flatten())
model.add(layers.Dense(64, activation='relu'))
model.add(layers.Dense(10))
```

Step 6: Compile and Train the Model

```
# Compile the model
model.compile(optimizer='adam',
```

```
loss=tf.keras.losses.SparseCategoricalCrossentropy(from_lo
gits=True),

        metrics=['accuracy'])


# Train the model

history = model.fit(train_images, train_labels, epochs=10,

            validation_data=(test_images, test_labels))
```

Step 7: Evaluate the Model

```
# Evaluate the model

test_loss,    test_acc    =    model.evaluate(test_images,
test_labels, verbose=2)

print(f'\nTest accuracy: {test_acc}')
```

Step 8: Visualize Training Results

```
# Visualize training results

plt.plot(history.history['accuracy'], label='accuracy')

plt.plot(history.history['val_accuracy'], label = 'val_accuracy')

plt.xlabel('Epoch')

plt.ylabel('Accuracy')

plt.ylim([0, 1])

plt.legend(loc='lower right')

plt.show()
```

How to Run the Project:

Save the above code in a Python file (e.g., **cnn_image_classification.py**).

Open a terminal and navigate to the directory containing the file.

Run the script using **python cnn_image_classification.py**.

This project demonstrates building a CNN using TensorFlow for image classification on the CIFAR-10 dataset. You can experiment with different CNN architectures, hyperparameters, and data augmentation techniques to improve performance.

8.3 Transfer Learning with TensorFlow Hub

Transfer learning allows leveraging pre-trained models for specific tasks. TensorFlow Hub provides a hub for reusable machine learning modules.

```
import tensorflow as tf

import tensorflow_hub as hub


# Load pre-trained model from TensorFlow Hub

model_url                    =                "https://tfhub.dev/google/tf2-preview/mobilenet_v2/classification/4"

feature_extractor = hub.KerasLayer(model_url, input_shape=(224, 224, 3))
```

8.3.1 Project: Transfer Learning with TensorFlow Hub for Text Classification

Transfer learning is a powerful technique in deep learning that leverages pre-trained models to solve a different but related problem. In this Python project, we'll use TensorFlow Hub for transfer learning on a text classification task.

Step 1: Install Dependencies

```
pip install tensorflow tensorflow-hub
```

Step 2: Import Libraries

```
import tensorflow as tf

import tensorflow_hub as hub

from tensorflow.keras.models import Sequential
```

```
from tensorflow.keras.layers import Dense
```

Step 3: Load and Preprocess Data

For this example, we'll use the IMDb movie reviews dataset for sentiment analysis.

```
# Load IMDb dataset

imdb = tf.keras.datasets.imdb

(train_data, train_labels), (test_data, test_labels) = imdb.load_data(num_words=10000)


# Preprocess the data

max_len = 128

train_data = tf.keras.preprocessing.sequence.pad_sequences(train_data, maxlen=max_len)

test_data = tf.keras.preprocessing.sequence.pad_sequences(test_data, maxlen=max_len)
```

Step 4: Load Pre-trained TensorFlow Hub Model

```
# Load pre-trained TensorFlow Hub model

embedding_url = "https://tfhub.dev/google/tf2-preview/gnews-swivel-20dim/1"

hub_layer = hub.KerasLayer(embedding_url, input_shape=[], dtype=tf.string, trainable=True)
```

Step 5: Build and Train the Model

```
# Build and train the model

model = Sequential([

    hub_layer,

    Dense(16, activation='relu'),

    Dense(1, activation='sigmoid')

])
```

```
# Compile the model

model.compile(optimizer='adam', loss='binary_crossentropy',
metrics=['accuracy'])


# Train the model

model.fit(train_data,          train_labels,          epochs=5,
validation_data=(test_data, test_labels))
```

Step 6: Evaluate the Model

```
# Evaluate the model

test_loss, test_acc = model.evaluate(test_data, test_labels,
verbose=2)

print(f'Test accuracy: {test_acc}')
```

How to Run the Project:

Save the above code in a Python file (e.g., **transfer_learning_text_classification.py**).

Open a terminal and navigate to the directory containing the file.

Run the script using **python transfer_learning_text_classification.py**.

This project demonstrates transfer learning using TensorFlow Hub for text classification. You can experiment with different pre-trained models from TensorFlow Hub and fine-tune the architecture based on your specific NLP task. Transfer learning is beneficial when dealing with limited labeled data, as it allows leveraging knowledge gained from large pre-trained models.

8.4 Object Detection with TensorFlow Object Detection API

TensorFlow Object Detection API simplifies the implementation of object detection models. It offers a range of pre-trained models for quick deployment.

```
import tensorflow as tf

from object_detection.utils import visualization_utils as viz_utils


# Example object detection using a pre-trained model

model = tf.saved_model.load('path/to/saved_model')

image = tf.image.decode_image(tf.io.read_file('path/to/image.jpg'))

input_tensor = tf.convert_to_tensor(image)

input_tensor = input_tensor[tf.newaxis, ...]


detections = model(input_tensor)


# Visualize the detections

viz_utils.visualize_boxes_and_labels_on_image_array(

    image.numpy(),

    detections['detection_boxes'][0].numpy(),

    detections['detection_classes'][0].numpy().astype(int),

    detections['detection_scores'][0].numpy(),

    category_index={1: {'id': 1, 'name': 'object'}},

    use_normalized_coordinates=True,

    max_boxes_to_draw=200,

    min_score_thresh=0.3,

    agnostic_mode=False

)
```

8.4.1 Project: Object Detection with TensorFlow Object Detection API

Object Detection is a crucial computer vision task where the goal is to identify and locate objects in an image or video. The TensorFlow Object Detection API provides pre-trained models and tools for building custom object detection models.

In this project, we'll use the TensorFlow Object Detection API to perform object detection on images.

Step 1: Install Dependencies

pip install tensorflow matplotlib

Step 2: Import Libraries

import os

import cv2

import numpy as np

import tensorflow as tf

from object_detection.utils import label_map_util

from object_detection.utils import visualization_utils as vis_util

import matplotlib.pyplot as plt

Step 3: Load Pre-trained Model and Labels

Download a pre-trained model and label map from the TensorFlow Model Zoo.

Path to the frozen detection graph and label map

PATH_TO_MODEL   = 'path/to/your/frozen_inference_graph.pb'

PATH_TO_LABELS = 'path/to/your/label_map.pbtxt'

Load the frozen detection graph

detection_graph = tf.compat.v1.Graph()

with detection_graph.as_default():

    od_graph_def = tf.compat.v1.GraphDef()

    with tf.io.gfile.GFile(PATH_TO_MODEL, 'rb') as fid:

        serialized_graph = fid.read()

        od_graph_def.ParseFromString(serialized_graph)

        tf.import_graph_def(od_graph_def, name='')

Load label map

```
category_index                              =
label_map_util.create_category_index_from_labelmap(PATH
_TO_LABELS, use_display_name=True)
```

Step 4: Load and Preprocess Image

Path to your test image

```
PATH_TO_IMAGE = 'path/to/your/test_image.jpg'
```


Load and preprocess the image

```
image = cv2.imread(PATH_TO_IMAGE)

image = cv2.cvtColor(image, cv2.COLOR_BGR2RGB)

image_np = np.expand_dims(image, axis=0)
```

Step 5: Perform Object Detection

Perform object detection

```
with detection_graph.as_default():

    with    tf.compat.v1.Session(graph=detection_graph)    as
sess:

        # Get handles to input and output tensors

        ops = tf.compat.v1.get_default_graph().get_operations()

        all_tensor_names = {output.name for op in ops for output
in op.outputs}

        tensor_dict = {}

        for key in ['num_detections', 'detection_boxes', 'detection_scores',
'detection_classes', 'detection_masks']:

            tensor_name = key + ':0'

            if tensor_name in all_tensor_names:

                tensor_dict[key]                              =
tf.compat.v1.get_default_graph().get_tensor_by_name(tensor_name)


        # Run inference
```

```
    output_dict                    =                    sess.run(tensor_dict,
feed_dict={tf.compat.v1.get_default_graph().get_tensor_by_name('image_te
nsor:0'): image_np})
```

Step 6: Visualize the Results

```
# Visualize the results

vis_util.visualize_boxes_and_labels_on_image_array(

    image,

    np.squeeze(output_dict['detection_boxes']),

    np.squeeze(output_dict['detection_classes']).astype(np.int32),

    np.squeeze(output_dict['detection_scores']),

    category_index,

    instance_masks=output_dict.get('detection_masks_reframed', None),

    use_normalized_coordinates=True,

    line_thickness=8

)


# Display the image with bounding boxes

plt.figure(figsize=(10, 6))

plt.imshow(image)

plt.show()
```

How to Run the Project:

Save the above code in a Python file (e.g., **object_detection_tf_api.py**).

Open a terminal and navigate to the directory containing the file.

Run the script using **python object_detection_tf_api.py**.

This project demonstrates how to use the TensorFlow Object Detection API to perform object detection on a test image. You can customize the project by choosing a different pre-trained model or fine-tuning on your specific dataset for more accurate results.

73

8.5 Image Segmentation with TensorFlow

Image segmentation involves dividing an image into segments for detailed analysis. TensorFlow provides tools for building segmentation models.

```
import tensorflow as tf

from tensorflow.keras import layers, models


# Example image segmentation model
model = models.Sequential([

    layers.Conv2D(64, (3, 3), activation='relu', input_shape=(256, 256, 3)),

    layers.Conv2D(64, (3, 3), activation='relu'),

    layers.MaxPooling2D((2, 2)),

    layers.Conv2D(128, (3, 3), activation='relu'),

    layers.Conv2D(128, (3, 3), activation='relu'),

    layers.MaxPooling2D((2, 2)),

    layers.Conv2D(256, (3, 3), activation='relu'),

    layers.Conv2D(256, (3, 3), activation='relu'),

    layers.Conv2D(256, (3, 3), activation='relu'),

    layers.MaxPooling2D((2, 2)),

    layers.Flatten(),

    layers.Dense(256, activation='relu'),

    layers.Dense(256, activation='softmax')

])
```

8.5.1 Project: Image Segmentation with TensorFlow Hub

Image segmentation is a computer vision task where the goal is to partition an image into meaningful segments. In this project, we'll use TensorFlow to perform image segmentation using a pre-trained model from the TensorFlow Hub.

Step 1: Install Dependencies

74

```
pip install tensorflow tensorflow-hub matplotlib
```

Step 2: Import Libraries

```
import tensorflow as tf

import tensorflow_hub as hub

import matplotlib.pyplot as plt
```

Step 3: Load Pre-trained Model

```
# Load pre-trained segmentation model from TensorFlow Hub

model_url = "https://tfhub.dev/tensorflow/tfjs-model/deeplab/coco-ssd/1"

model = hub.load(model_url)
```

Step 4: Load and Preprocess Image

```
# Path to your test image

PATH_TO_IMAGE = 'path/to/your/test_image.jpg'


# Load and preprocess the image

image = tf.io.read_file(PATH_TO_IMAGE)

image = tf.image.decode_jpeg(image)

image = tf.image.convert_image_dtype(image, dtype=tf.uint8)

image = tf.expand_dims(image, 0)
```

Step 5: Perform Image Segmentation

```
# Perform image segmentation

segmentation_result = model(image)
```

Step 6: Visualize the Results

```
# Extract segmentation masks and class labels

masks = segmentation_result['segmentation_mask']

class_ids = segmentation_result['class_ids']


# Visualize the results

fig, axes = plt.subplots(1, 2, figsize=(12, 6))


# Display the original image
```

```
axes[0].imshow(image[0])

axes[0].axis('off')

axes[0].set_title('Original Image')


# Display the segmented image

segmentation_mask = tf.argmax(masks, axis=-1)

segmentation_mask = tf.expand_dims(segmentation_mask, axis=-1)

segmentation_mask = tf.image.resize(segmentation_mask, (image.shape[1],
image.shape[2]))

segmentation_mask = tf.cast(segmentation_mask, dtype=tf.uint8)

axes[1].imshow(segmentation_mask[0, :, :, 0], cmap='viridis', alpha=0.7)

axes[1].imshow(image[0], alpha=0.4)

axes[1].axis('off')

axes[1].set_title('Segmentation Result')


plt.show()
```

How to Run the Project:

Save the above code in a Python file (e.g., **image_segmentation_tf_hub.py**).

Open a terminal and navigate to the directory containing the file.

Run the script using **python image_segmentation_tf_hub.py**.

This project demonstrates image segmentation using a pre-trained DeepLab model from TensorFlow Hub. You can experiment with different models from TensorFlow Hub or use custom segmentation models for your specific application.

8.6 TensorFlow Lite for Edge Devices

TensorFlow Lite enables the deployment of models on edge devices, allowing for real-time inference on devices with limited resources.

```
import tensorflow as tf

converter                                              =
tf.lite.TFLiteConverter.from_saved_model('path/to/saved_model')

tflite_model = converter.convert()


# Save the TensorFlow Lite model to a file

with open('model.tflite', 'wb') as f:

    f.write(tflite_model)
```

8.6.1 Project: TensorFlow Lite for Image Classification on Raspberry Pi

TensorFlow Lite is designed for running machine learning models on edge devices with limited computational resources. In this project, we'll use TensorFlow Lite to deploy a simple image classification model on a Raspberry Pi, a popular edge device.

Step 1: Train a Simple Image Classification Model

Train a simple image classification model using TensorFlow. Save the model in TensorFlow SavedModel format.

Step 2: Convert the Model to TensorFlow Lite format

Use the TensorFlow Lite Converter to convert the SavedModel to TensorFlow Lite format.

```
tflite_convert   --saved_model_dir=path/to/saved_model   --output_file=model.tflite
```

Step 3: Set Up the Raspberry Pi

Make sure your Raspberry Pi is set up with Raspbian OS, and it's connected to the internet.

Step 4: Install TensorFlow Lite Interpreter on Raspberry Pi

pip install https://dl.google.com/coral/python/tflite_runtime-2.5.0.post1-cp37-cp37m-linux_armv7l.whl

Step 5: Write Python Script for Inference on Raspberry Pi

Create a Python script (**inference_script.py**) on your Raspberry Pi to load the TensorFlow Lite model and perform inference on a sample image.

```python
import cv2

import numpy as np

from tflite_runtime.interpreter import Interpreter


# Load TensorFlow Lite model

interpreter = Interpreter(model_path="model.tflite")

interpreter.allocate_tensors()

input_details = interpreter.get_input_details()

output_details = interpreter.get_output_details()


# Preprocess input image

image_path = "path/to/your/sample_image.jpg"

image = cv2.imread(image_path)

input_image = cv2.resize(image, (input_details[0]['shape'][2],
input_details[0]['shape'][1]))

input_image = np.expand_dims(input_image, axis=0)

input_image = input_image.astype(np.float32) / 255.0


# Set input tensor

interpreter.set_tensor(input_details[0]['index'], input_image)


# Run inference

interpreter.invoke()
```

78

```
# Get the output tensor

output = interpreter.get_tensor(output_details[0]['index'])

predicted_class = np.argmax(output[0])


# Print the predicted class

print(f"Predicted Class: {predicted_class}")
```

Step 6: Run the Inference Script on Raspberry Pi

Transfer the script and the sample image to your Raspberry Pi and run the script.

```
python inference_script.py
```

Note:

Make sure to replace **"path/to/saved_model"** with the actual path to your TensorFlow SavedModel.

Ensure that the TensorFlow Lite model is compatible with the hardware architecture of the Raspberry Pi.

Adjust the script according to your model's input and output details.

This project demonstrates deploying a TensorFlow Lite model on a Raspberry Pi for image classification. You can adapt this example for other edge devices or modify the script for different types of models.

Chapter 9: Deployment and Production with TensorFlow

Deploying machine learning models into production is a crucial step in turning research into practical solutions. TensorFlow offers various tools and techniques for deploying models effectively and ensuring they perform well in real-world scenarios. In this chapter, we'll explore strategies for deploying and managing TensorFlow models in production environments.

9.1 TensorFlow Serving

TensorFlow Serving is a flexible, high-performance serving system for machine learning models. It allows you to deploy and serve your trained models using a simple API.

```
# Example serving command for TensorFlow Serving

docker    run    -p    8501:8501    --name=tf_model_serving    --mount
type=bind,source=/path/to/saved_model,target=/models/model        -e
MODEL_NAME=model -t tensorflow/serving
```

With TensorFlow Serving, you can expose your model via REST APIs, gRPC, or other standard protocols. This makes it easy to integrate your model into various applications and services.

9.1.1 Project: TensorFlow Serving for Model Deployment

TensorFlow Serving is an open-source serving system for machine learning models designed for production environments. It allows you to deploy machine learning models using a flexible, high-performance serving system, making it easier to manage and scale your models. In this project, we'll set up TensorFlow Serving and deploy a pre-trained model for serving.

Step 1: Install Docker

If you don't have Docker installed, you can install it from the official Docker website: Install Docker.

Step 2: Pull TensorFlow Serving Docker Image

Pull the TensorFlow Serving Docker image from the official repository:

docker pull tensorflow/serving

Step 3: Convert the Model to TensorFlow SavedModel format

Make sure you have a pre-trained model and convert it to the TensorFlow SavedModel format. For example:

import tensorflow as tf

from tensorflow.keras.applications import MobileNetV2

Load pre-trained MobileNetV2 model

model = MobileNetV2(weights='imagenet')

model.save("saved_model/mobilenetv2/1/")

Step 4: Run TensorFlow Serving Docker Container

Run TensorFlow Serving as a Docker container, exposing the necessary ports:

docker run -p 8501:8501 --name=tf-serving --mount type=bind,source=$(pwd)/saved_model,target=/models/mobi lenetv2 -e MODEL_NAME=mobilenetv2 -t tensorflow/serving

Step 5: Send Requests to the TensorFlow Serving API

You can now send image classification requests to the TensorFlow Serving API. For example:

import requests

import json

import cv2

import numpy as np

Load and preprocess an image

image_path = "path/to/your/image.jpg"

image = cv2.imread(image_path)

image = cv2.resize(image, (224, 224))

image = image.astype(np.float32) / 255.0

image = np.expand_dims(image, axis=0)

Prepare the JSON request

data = json.dumps({"signature_name": "serving_default", "instances": image.tolist()})

Send the request to the TensorFlow Serving API

headers = {"content-type": "application/json"}

url = "http://localhost:8501/v1/models/mobilenetv2:predict"

response = requests.post(url, data=data, headers=headers)

Parse the response

predictions = json.loads(response.text)["predictions"]

Print the predictions

print(predictions)

How to Run the Project:

Follow the steps above to set up TensorFlow Serving using Docker.

Convert your pre-trained model to the TensorFlow SavedModel format.

Run the TensorFlow Serving Docker container.

Send requests to the TensorFlow Serving API using the provided Python script.

This project demonstrates the process of setting up TensorFlow Serving using Docker and deploying a pre-trained model for serving. You can adapt this example to deploy different models or modify the script for different types of inference tasks.

9.2 TensorFlow Lite for deploying machine learning models on Edge Devices

TensorFlow Lite is designed for deploying machine learning models on edge devices, making it ideal for applications where real-time inference is required on devices with limited resources.

```
import tensorflow as tf

converter                                                    =
tf.lite.TFLiteConverter.from_saved_model('path/to/saved_model')

tflite_model = converter.convert()


# Save the TensorFlow Lite model to a file

with open('model.tflite', 'wb') as f:

    f.write(tflite_model)
```

TensorFlow Lite supports various platforms, including Android and iOS, enabling the integration of machine learning models into mobile applications.

9.2.1 Project: TensorFlow Lite Model Deployment on Raspberry Pi

Deploying machine learning models on edge devices using TensorFlow Lite is a common use case. In this project, we'll create a simple example of deploying a TensorFlow Lite image classification model on a Raspberry Pi.

Step 1: Train and Convert a Simple Image Classification Model

Train a simple image classification model using TensorFlow and convert it to TensorFlow Lite format.

```python
import tensorflow as tf

from tensorflow.keras.models import Sequential

from tensorflow.keras.layers import Dense, Flatten


# Train a simple model
model = Sequential([

    Flatten(input_shape=(224, 224, 3)),

    Dense(128, activation='relu'),

    Dense(10, activation='softmax')

])


# Compile and train the model
# ...


# Convert the model to TensorFlow Lite format
converter = tf.lite.TFLiteConverter.from_keras_model(model)

tflite_model = converter.convert()
```

```
with open('model.tflite', 'wb') as f:

    f.write(tflite_model)
```

Step 2: Set Up Raspberry Pi

Ensure your Raspberry Pi is set up and connected to the internet.

Step 3: Install TensorFlow Lite Interpreter on Raspberry Pi

```
pip install tflite_runtime
```

Step 4: Write Python Script for Inference on Raspberry Pi

Create a Python script (**inference_script.py**) on your Raspberry Pi to load the TensorFlow Lite model and perform inference on a sample image.

```
import cv2

import numpy as np

from tflite_runtime.interpreter import Interpreter


# Load TensorFlow Lite model

interpreter = Interpreter(model_path="model.tflite")

interpreter.allocate_tensors()

input_details = interpreter.get_input_details()

output_details = interpreter.get_output_details()


# Preprocess input image

image_path = "path/to/your/sample_image.jpg"

image = cv2.imread(image_path)

input_image            =            cv2.resize(image,
(input_details[0]['shape'][2],
input_details[0]['shape'][1]))
```

```
input_image    =    np.expand_dims(input_image,
axis=0)
```

```
input_image = input_image.astype(np.float32) / 255.0
```

```
# Set input tensor
```

```
interpreter.set_tensor(input_details[0]['index'], input_image)
```

```
# Run inference
```

```
interpreter.invoke()
```

```
# Get the output tensor
```

```
output = interpreter.get_tensor(output_details[0]['index'])
```

```
predicted_class = np.argmax(output[0])
```

```
# Print the predicted class
```

```
print(f"Predicted Class: {predicted_class}")
```

Step 5: Run the Inference Script on Raspberry Pi

Transfer the script and the sample image to your Raspberry Pi and run the script.

```
python inference_script.py
```

Note:

Ensure that the TensorFlow Lite model is compatible with the hardware architecture of the Raspberry Pi.

Adjust the script according to your model's input and output details.

This project demonstrates deploying a TensorFlow Lite model on a Raspberry Pi for image classification. You can adapt this example for other edge devices or modify the script for different types of models.

9.3 TensorFlow Extended (TFX)

TensorFlow Extended (TFX) is an end-to-end platform for deploying production-ready machine learning models. TFX includes components for data validation, model analysis, and continuous evaluation, providing a comprehensive solution for managing the machine learning lifecycle.

```
# Example TFX pipeline

# (Note: TFX pipeline setup is beyond the scope of a code snippet)

import tfx


pipeline = tfx.dsl.Pipeline(

    ... # Define your TFX components and pipeline steps here

)


# Run the TFX pipeline

tfx.orchestration.LocalDagRunner().run(pipeline)
```

TFX is designed to scale seamlessly from local development to large-scale production deployments.

9.3.1 Project: TensorFlow Extended (TFX) Pipeline

TensorFlow Extended (TFX) is an end-to-end platform for deploying production-ready machine learning models. It facilitates the development and deployment of scalable and maintainable production pipelines. In this project, we'll set up a simple TFX pipeline for training and serving a TensorFlow model.

Step 1: Install Required Libraries

pip install tfx

Step 2: Set Up TFX Project Structure

Create a TFX project structure with the following components:

pipeline: Define your TFX pipeline.

components: Implement custom TFX components (e.g., Trainer, Tuner).

modules: Define feature transformations.

serving_model: Store the TensorFlow SavedModel for serving.

Step 3: Define TFX Pipeline

Create a **pipeline.py** file to define your TFX pipeline. Here's a basic example:

```
# pipeline.py

from        tfx.dsl.components.common.importer        import
ImportExampleGen

from        tfx.dsl.components.common.statistics_gen        import
StatisticsGen

from  tfx.dsl.components.common.example_validator  import
ExampleValidator

from tfx.dsl.components.common.transform import Transform

from tfx.dsl.components.common.trainer import Trainer

from tfx.dsl.components.common.tuner import Tuner

from tfx.dsl.components.common.evaluator import Evaluator

from        tfx.dsl.components.common.infra_validator        import
InfraValidator

from        tfx.dsl.components.common.bulk_inferrer        import
BulkInferrer

from tfx.orchestration import data_types

from tfx.types import standard_artifacts

from tfx.types import channel_utils

from tfx.utils.dsl_utils import external_input

from tfx.proto import example_gen_pb2

from tfx.proto import infra_validator_pb2


# Define pipeline components
```

```python
# ExampleGen
examples = external_input("path/to/dataset")
example_gen = ImportExampleGen(input=examples)


# StatisticsGen
statistics_gen                                          =
StatisticsGen(examples=example_gen.outputs['examples'])


# ExampleValidator
example_validator = ExampleValidator(
    statistics=statistics_gen.outputs['statistics'],
    examples=example_gen.outputs['examples']
)


# Transform
transform = Transform(
    examples=example_gen.outputs['examples'],
    schema=example_validator.outputs['output'],
)


# Trainer
trainer = Trainer(
    module_file="path/to/module.py",

transformed_examples=transform.outputs['transformed_exa
mples'],
    schema=example_validator.outputs['output'],
```

```
)

# Tuner (Optional)
tuner = Tuner(
    module_file="path/to/module.py",
    examples=example_gen.outputs['examples'],
    transform_graph=transform.outputs['transform_graph'],
    schema=example_validator.outputs['output'],
)


# Evaluator
evaluator = Evaluator(
    examples=example_gen.outputs['examples'],
    model=trainer.outputs['model'],
)


# InfraValidator (Optional)
infra_validator = InfraValidator(
    model=trainer.outputs['model'],
    infra_blessing=evaluator.outputs['blessing'],
    serving_spec=infra_validator_pb2.ServingSpec(),
)


# BulkInferrer (Optional)
bulk_inferrer = BulkInferrer(
    examples=example_gen.outputs['examples'],
    model=trainer.outputs['model'],
)
```

```
# Define the TFX pipeline
components = [
    example_gen,
    statistics_gen,
    example_validator,
    transform,
    trainer,
    tuner,
    evaluator,
    infra_validator,
    bulk_inferrer,
]


pipeline = data_types.Pipeline(
    pipeline_name="my_tfx_pipeline",
    pipeline_root="path/to/pipeline_root",
    components=components,
    enable_cache=True,
)
```

Step 4: Create TensorFlow Model

Create a TensorFlow model file, e.g., **module.py**, for training and serving.

```
# module.py
import tensorflow as tf
from tensorflow import keras
from tensorflow.keras import layers
```

```
def create_model():

    model = keras.Sequential([

        layers.Input(shape=(...)),  # Define input shape

        # Add layers as needed

        layers.Dense(64, activation='relu'),

        layers.Dense(10, activation='softmax')

    ])

    model.compile(optimizer='adam',
loss='sparse_categorical_crossentropy', metrics=['accuracy'])

    return model
```

Step 5: Run the TFX Pipeline

Run the TFX pipeline using the TFX CLI.

```
tfx  pipeline  create  --pipeline-path=path/to/pipeline.py  --
endpoint=localhost:8500

tfx  run  create  --pipeline-name=my_tfx_pipeline  --
endpoint=localhost:8500
```

This project provides a basic example of setting up a TFX pipeline for training and serving a TensorFlow model. Customize the pipeline and components based on your specific requirements and data.

9.4 TensorFlow Model Optimization

Optimizing your models for deployment is crucial for achieving efficient and fast inference. TensorFlow Model Optimization provides tools for optimizing and quantizing models, reducing their size and making them more suitable for deployment on edge devices.

```
import tensorflow as tf
```

```
from tensorflow_model_optimization.python.core.quantization.keras import
quantize_model
```

```
# Example model quantization
```

```
quantized_model = quantize_model(model)
```

9.4.1 Project: TensorFlow Model Optimization for Mobile Deployment

TensorFlow Model Optimization (TFLite) allows you to optimize and quantize models for deployment on edge devices. In this project, we'll create a simple example of optimizing a TensorFlow model for deployment on a mobile device using quantization.

Step 1: Train a Simple TensorFlow Model

Train a simple TensorFlow model. Save the trained model in the SavedModel format.

```
import tensorflow as tf
```

```
from tensorflow.keras.models import Sequential
```

```
from tensorflow.keras.layers import Dense
```

```
# Create a simple model
```

```
model = Sequential([
    Dense(128, activation='relu', input_shape=(28, 28)),
    Dense(10, activation='softmax')
])
```

```
# Compile and train the model
```

```
# ...
```

```
# Save the model in SavedModel format
```

```
model.save("path/to/saved_model")
```

Step 2: Optimize the Model using TensorFlow Model Optimization

Use TensorFlow Model Optimization to optimize and quantize the model.

```python
import tensorflow as tf

from tensorflow_model_optimization.python.core.sparsity.keras import prune, pruning_callbacks


# Load the trained model

model = tf.keras.models.load_model("path/to/saved_model")


# Define a pruning schedule

pruning_schedule = tfmot.sparsity.keras.PolynomialDecay(

    initial_sparsity=0.0,

    final_sparsity=0.5,

    begin_step=0,

    end_step=1000

)


# Apply pruning to the model

model = prune.prune_low_magnitude(model, pruning_schedule=pruning_schedule)


# Compile the model

model.compile(optimizer='adam',

        loss='sparse_categorical_crossentropy',

        metrics=['accuracy'])


# Display the sparsity of the model
```

```
print("Sparsity                    after                    pruning:",
tfmot.sparsity.keras.prune_low_magnitude.get_sparsity(mod
el))
```

Step 3: Fine-Tune the Pruned Model

Fine-tune the pruned model to recover any accuracy loss.

Fine-tune the pruned model

...

Step 4: Convert the Model to TensorFlow Lite Format

Convert the fine-tuned model to TensorFlow Lite format for deployment on mobile devices.

converter = tf.lite.TFLiteConverter.from_keras_model(model)

tflite_model = converter.convert()

with open("optimized_model.tflite", "wb") as f:

   f.write(tflite_model)

How to Run the Project:

Train a simple TensorFlow model and save it in the SavedModel format.

Optimize and quantize the model using TensorFlow Model Optimization.

Fine-tune the pruned model if needed.

Convert the model to TensorFlow Lite format for deployment on mobile devices.

This project demonstrates how to use TensorFlow Model Optimization to optimize and quantize a TensorFlow model for deployment on edge devices. Customize the pruning schedule and other parameters based on your specific requirements.

9.5 Model Versioning and Monitoring

In a production environment, managing different versions of your machine learning models is essential. Tools like TensorFlow Model Analysis and TensorBoard can help monitor and compare model performance over time.

```
# Example TensorBoard command for monitoring model training

tensorboard --logdir=/path/to/logs
```

TensorFlow Model Analysis allows you to analyze and visualize model performance on different datasets, aiding in making informed decisions about model updates.

9.5.1 Project: TensorFlow Model Analysis

TensorFlow Model Analysis (TFMA) is a powerful tool for evaluating and analyzing machine learning models. It provides metrics, visualizations, and tools for understanding model performance. In this project, we'll create a simple example of using TFMA to analyze and evaluate a TensorFlow model.

Step 1: Install Required Libraries

```
pip install tensorflow tensorflow_model_analysis
```

Step 2: Train a TensorFlow Model

Train a TensorFlow model using your dataset. Save the trained model in the SavedModel format.

```
import tensorflow as tf

from tensorflow.keras.models import Sequential

from tensorflow.keras.layers import Dense


# Create a simple model
model = Sequential([

    Dense(128, activation='relu', input_shape=(28, 28)),

    Dense(10, activation='softmax')
```

```
])
```

Compile and train the model

```
# ...
```

Save the model in SavedModel format

```
model.save("path/to/saved_model")
```

Step 3: Evaluate the Model

Evaluate the model on a test dataset.

Load the trained model

```
model = tf.keras.models.load_model("path/to/saved_model")
```

Evaluate the model on a test dataset

```
# ...
```

Step 4: Create TFMA Metrics and Slices

Create TFMA metrics and slices for detailed model analysis.

```
import tensorflow_model_analysis as tfma
```

Create TFMA metrics

```
metrics_specs = tfma.metrics.specs_from_metrics({
    'accuracy':
tfma.metrics.MeanMetricWrapper(tf.keras.metrics.SparseCat
egoricalAccuracy()),
    'precision': tfma.metrics.Precision(),
    'recall': tfma.metrics.Recall(),
    'confusion_matrix': tfma.metrics.ConfusionMatrixPlot(),
})
```

```
# Create slices

slices = [

    tfma.slicer.SingleSliceSpec(),  # Overall metrics

    tfma.slicer.SingleSliceSpec(columns=['feature_column']),
# Specific feature

    tfma.slicer.SingleSliceSpec(features=[('feature_column',
1)]),  # Specific feature value

]
```

Step 5: Run TFMA Analysis

Run TFMA analysis on the evaluated model.

```
# Create a TFMA EvalSavedModel

eval_saved_model_path = "path/to/eval_saved_model"

tfma.export.export_eval_savedmodel(

    estimator=None,  # Use None if not using Estimator API

    export_dir_base=eval_saved_model_path,


eval_input_receiver_fn=tfma.export.build_parsing_eval_input
_receiver_fn(tfma.types.FeatureSpecType.VAR_LEN_FEAT
URE),


serving_input_receiver_fn=tfma.export.build_parsing_servin
g_input_receiver_fn(tfma.types.FeatureSpecType.VAR_LEN
_FEATURE),

    model=model,

)


# Create an EvalConfig

eval_config = tfma.EvalConfig(

    model_specs=[

      tfma.ModelSpec(

        label_key='label',
```

```
        prediction_key='dense_1',
    )
],
slicing_specs=slices,
metrics_specs=metrics_specs,
)


# Run TFMA analysis
eval_result = tfma.run_model_analysis(
    eval_shared_model=tfma.default_eval_shared_model(
        eval_saved_model_path=eval_saved_model_path,
        add_metrics_callbacks=[],
    ),
    eval_config=eval_config,
    data_location="path/to/test_dataset.tfrecords",
    output_path="path/to/tfma_output",
    file_format="tfrecords",
)
```

How to Run the Project:

Train a TensorFlow model and save it in the SavedModel format.

Evaluate the model on a test dataset.

Create TFMA metrics and slices.

Run TFMA analysis on the evaluated model.

This project provides a basic example of using TensorFlow Model Analysis (TFMA) for evaluating and analyzing a TensorFlow model. Customize the metrics, slices, and other configurations based on your specific requirements.

9.6 TensorFlow Enterprise

For large-scale deployments and enterprise-grade solutions, TensorFlow Enterprise provides additional features, support, and services. It is designed to meet the requirements of businesses that rely on machine learning at scale.

9.7 Summary

Deploying machine learning models into production with TensorFlow involves selecting the right tools and strategies to meet your specific needs. Whether you're serving models via TensorFlow Serving, deploying on edge devices with TensorFlow Lite, managing the entire machine learning lifecycle with TFX, optimizing models for efficiency, or monitoring model performance, TensorFlow offers a comprehensive set of solutions. As you navigate the deployment and production landscape, consider the specific requirements of your application and leverage TensorFlow's versatility to ensure a seamless transition from development to real-world deployment.

Chapter 10: TensorFlow for Reinforcement Learning

Reinforcement Learning (RL) is a paradigm of machine learning where agents learn to make decisions by interacting with an environment. TensorFlow provides a versatile platform for developing and implementing Reinforcement Learning models. In this chapter, we'll explore how TensorFlow can be used to build, train, and deploy RL models.

10.1 OpenAI Gym Integration

OpenAI Gym is a toolkit for developing and comparing reinforcement learning algorithms. TensorFlow can seamlessly integrate with OpenAI Gym to create custom environments and interact with them using RL models.

```
import gym

import tensorflow as tf


# Create a custom Gym environment

class CustomEnvironment(gym.Env):
```

```
def __init__(self):

    super(CustomEnvironment, self).__init__()
    # Define your environment parameters and spaces


def step(self, action):
    # Define the logic for the environment's dynamics
    pass


def reset(self):
    # Reset the environment to its initial state
    pass


def render(self, mode='human'):
    # Render the environment for visualization
    pass


# Use TensorFlow to interact with the custom environment
env = CustomEnvironment()
state = env.reset()
action = tf.argmax(model(state), axis=1).numpy()
next_state, reward, done, _ = env.step(action)
```

10.1.1 Project: OpenAI Gym Integration using TensorFlow

Integrating OpenAI Gym with TensorFlow allows you to train and evaluate reinforcement learning models in various environments. In this project, we'll create a simple example of integrating OpenAI Gym with TensorFlow to train a basic reinforcement learning agent.

Step 1: Install Required Libraries

pip install gym tensorflow

Step 2: Create a Simple OpenAI Gym Environment

Create a simple OpenAI Gym environment. In this example, we'll use the CartPole environment.

```
import gym
```

```
# Create the CartPole environment
env = gym.make('CartPole-v1')
```

```
# Get the action space and observation space
action_space = env.action_space.n
observation_space = env.observation_space.shape[0]
```

Step 3: Build a Simple Neural Network for the Agent

Build a simple neural network using TensorFlow to serve as the Q-function approximator for the reinforcement learning agent.

```
import tensorflow as tf
from tensorflow.keras import layers
```

```
# Define a simple neural network model
model = tf.keras.Sequential([
    layers.Dense(64,                          activation='relu',
input_shape=(observation_space,)),
    layers.Dense(32, activation='relu'),
    layers.Dense(action_space, activation='linear')
])
```

Step 4: Define the Reinforcement Learning Agent

Define the reinforcement learning agent using TensorFlow. Use an epsilon-greedy strategy for exploration.

```
import numpy as np
```

```python
class EpsilonGreedyAgent:

    def __init__(self, model, action_space, epsilon=0.1):
        self.model = model
        self.action_space = action_space
        self.epsilon = epsilon


    def choose_action(self, state):
        if np.random.rand() < self.epsilon:
            return np.random.choice(self.action_space)
        else:
            q_values = self.model.predict(state.reshape(1, -1))[0]
            return np.argmax(q_values)
```

Step 5: Train the Reinforcement Learning Agent

Train the reinforcement learning agent using Q-learning.

```python
import random


# Hyperparameters
learning_rate = 0.001
discount_factor = 0.99
epsilon_decay = 0.995
epsilon_min = 0.01
num_episodes = 1000


# Compile the model
model.compile(optimizer=tf.optimizers.Adam(learning_rate),
loss='mse')


# Initialize the agent
```

```
agent = EpsilonGreedyAgent(model, action_space)


# Training loop
for episode in range(num_episodes):
    state = env.reset()
    state = np.reshape(state, [1, observation_space])


    total_reward = 0


    while True:
        # Choose an action
        action = agent.choose_action(state)


        # Take the chosen action
        next_state, reward, done, _ = env.step(action)
        next_state = np.reshape(next_state, [1, observation_space])


        # Update Q-values using Q-learning
        target = reward + discount_factor * np.max(agent.model.predict(next_state)[0])
        q_values = agent.model.predict(state)
        q_values[0][action] = target
        agent.model.fit(state, q_values, epochs=1, verbose=0)


        total_reward += reward
        state = next_state
```

```
if done:

    break
```

```
# Decay epsilon

agent.epsilon = max(epsilon_min, agent.epsilon *
epsilon_decay)
```

```
# Print the total reward for the episode

print(f"Episode {episode + 1}/{num_episodes}, Total
Reward: {total_reward}")
```

```
# Close the environment

env.close()
```

How to Run the Project:

Install the required libraries (**gym** and **tensorflow**).

Create an OpenAI Gym environment (e.g., CartPole).

Build a simple neural network for the Q-function approximator.

Define the reinforcement learning agent.

Train the reinforcement learning agent using Q-learning.

This project provides a basic example of integrating OpenAI Gym with TensorFlow for training a reinforcement learning agent. Customize the neural network architecture, hyperparameters, and other components based on your specific requirements and the characteristics of your chosen environment.

10.2 TensorFlow Agents

TensorFlow Agents (TF-Agents) is a library for building Reinforcement Learning models using TensorFlow. It provides components and tools to create a wide range of RL algorithms.

```python
import tensorflow as tf

from tf_agents.networks import q_network

from tf_agents.agents.dqn import dqn_agent

from tf_agents.environments import suite_gym

from tf_agents.environments import tf_py_environment


# Create a TensorFlow environment
env = suite_gym.load('CartPole-v1')

env = tf_py_environment.TFPyEnvironment(env)


# Define the Q-network
fc_layer_params = (100, )

q_net = q_network.QNetwork(
    env.observation_spec(),
    env.action_spec(),
    fc_layer_params=fc_layer_params
)


# Instantiate the DQN agent
optimizer = tf.compat.v1.train.AdamOptimizer(learning_rate=1e-3)

train_step_counter = tf.Variable(0)

agent = dqn_agent.DqnAgent(
    env.time_step_spec(),
    env.action_spec(),
    q_network=q_net,
    optimizer=optimizer,
    td_errors_loss_fn=common.element_wise_squared_loss,
    train_step_counter=train_step_counter
)
```

agent.initialize()

10.2.1 Project: Building Reinforcement Learning Models using TensorFlow

Building a reinforcement learning model using TensorFlow involves creating a neural network and implementing the training loop to update the model based on the agent's interactions with the environment. Below is a simple example using TensorFlow and the OpenAI Gym CartPole environment.

Step 1: Install Required Libraries

pip install gym tensorflow

Step 2: Create a Simple Neural Network for the Agent

Build a simple neural network using TensorFlow to serve as the Q-function approximator for the reinforcement learning agent.

```
import tensorflow as tf

from tensorflow.keras import layers


# Define the Q-network

def build_q_network(input_shape, num_actions):

    model = tf.keras.Sequential([

        layers.Dense(64,                         activation='relu',
input_shape=input_shape),

        layers.Dense(32, activation='relu'),

        layers.Dense(num_actions, activation='linear')

    ])

    return model


# Create the Q-network

input_shape = (4,)  # CartPole state has 4 dimensions

num_actions = 2  # CartPole has 2 actions (left or right)
```

```
q_network = build_q_network(input_shape, num_actions)
```

Step 3: Define the Reinforcement Learning Agent

Define the reinforcement learning agent using TensorFlow. In this example, we'll use the DQN (Deep Q-Network) algorithm.

```
import numpy as np


class DQNAgent:

    def __init__(self, q_network, num_actions,
discount_factor=0.99, epsilon=1.0, epsilon_decay=0.995,
epsilon_min=0.01):

        self.q_network = q_network

        self.num_actions = num_actions

        self.discount_factor = discount_factor

        self.epsilon = epsilon

        self.epsilon_decay = epsilon_decay

        self.epsilon_min = epsilon_min


    def choose_action(self, state):
        if np.random.rand() < self.epsilon:
            return np.random.choice(self.num_actions)
        else:
            q_values = self.q_network.predict(state.reshape(1, -
1))[0]
            return np.argmax(q_values)


    def update_model(self, state, action, reward, next_state,
done):
        # Q-learning update
        target = reward + self.discount_factor *
np.max(self.q_network.predict(next_state.reshape(1, -1))[0])
```

```
q_values = self.q_network.predict(state.reshape(1, -1))

q_values[0][action] = target

self.q_network.fit(state.reshape(1,     -1),     q_values,
epochs=1, verbose=0)


    # Decay epsilon

self.epsilon  =  max(self.epsilon * self.epsilon_decay,
self.epsilon_min)
```

Step 4: Training Loop

Implement the training loop to train the DQN agent on the CartPole environment.

```
import gym


# Create the CartPole environment

env = gym.make('CartPole-v1')


# Hyperparameters

num_episodes = 500

batch_size = 32


# Initialize the agent

agent = DQNAgent(q_network, num_actions)


# Training loop

for episode in range(num_episodes):
    state = env.reset()
    state = state.reshape(1, -1)
    total_reward = 0
```

```
while True:
    # Choose an action
    action = agent.choose_action(state)


    # Take the chosen action
    next_state, reward, done, _ = env.step(action)
    next_state = next_state.reshape(1, -1)


    # Update the model
    agent.update_model(state, action, reward, next_state, done)


    total_reward += reward
    state = next_state


    if done:
        break


    # Print the total reward for the episode
    print(f"Episode {episode + 1}/{num_episodes}, Total Reward: {total_reward}")


# Close the environment
env.close()
```

How to Run the Project:

Install the required libraries (**gym** and **tensorflow**).

Build a simple neural network for the Q-function approximator.

Define the reinforcement learning agent using DQN.

Implement the training loop to train the agent on the CartPole environment.

This project provides a basic example of building a reinforcement learning model using TensorFlow and training it with the DQN algorithm. Customize the neural network architecture, hyperparameters, and other components based on your specific requirements and the characteristics of your chosen environment.

10.3 Proximal Policy Optimization (PPO) with TensorFlow

Proximal Policy Optimization (PPO) is a popular RL algorithm. TensorFlow can be used to implement PPO, providing a stable and effective approach for training agents.

```python
import tensorflow as tf

from tensorflow.keras.models import Sequential

from tensorflow.keras.layers import Dense

from tensorflow.keras.optimizers import Adam

import gym


# Create a PPO agent with TensorFlow

class PPOAgent:

    def __init__(self, state_dim, action_dim, action_bound):

        self.actor = self.build_actor(state_dim, action_dim, action_bound)

        self.critic = self.build_critic(state_dim)


    def build_actor(self, state_dim, action_dim, action_bound):

        model = Sequential([

            Dense(64, input_dim=state_dim, activation='relu'),

            Dense(32, activation='relu'),
```

```
    Dense(action_dim, activation='tanh')

])

model.compile(loss='mse', optimizer=Adam(learning_rate=0.001))

return model


 def build_critic(self, state_dim):

  model = Sequential([

    Dense(64, input_dim=state_dim, activation='relu'),

    Dense(32, activation='relu'),

    Dense(1)

  ])

  model.compile(loss='mse', optimizer=Adam(learning_rate=0.001))

  return model
```

```
# Instantiate and train the PPO agent

state_dim = 4   # Example state dimension for CartPole
environment

action_dim = 1  # Example action dimension for CartPole
environment

action_bound = 1.0  # Example action bound for CartPole
environment


ppo_agent    =    PPOAgent(state_dim,    action_dim,
action_bound)

# Train the agent using PPO algorithm
```

10.3.1 Project: Proximal Policy Optimization (PPO) with TensorFlow

Proximal Policy Optimization (PPO) is a popular algorithm in reinforcement learning. In this project, we'll implement PPO using TensorFlow and apply it to the OpenAI Gym environment. Note that PPO typically requires a more complex setup compared to simple DQN, as it involves training a policy network and a value network.

Step 1: Install Required Libraries

pip install gym tensorflow

Step 2: Create the Policy and Value Networks

Build neural networks for the policy and value functions.

import tensorflow as tf

from tensorflow.keras import layers


```python
class PolicyNetwork(tf.keras.Model):

    def __init__(self, num_actions):

        super(PolicyNetwork, self).__init__()

        self.dense1 = layers.Dense(64, activation='relu')

        self.dense2 = layers.Dense(64, activation='relu')

        self.policy_head    =    layers.Dense(num_actions,
activation='softmax')


    def call(self, state):

        x = self.dense1(state)

        x = self.dense2(x)

        return self.policy_head(x)


class ValueNetwork(tf.keras.Model):

    def __init__(self):

        super(ValueNetwork, self).__init__()

        self.dense1 = layers.Dense(64, activation='relu')

        self.dense2 = layers.Dense(64, activation='relu')

        self.value_head = layers.Dense(1, activation=None)
```

```
def call(self, state):

    x = self.dense1(state)

    x = self.dense2(x)

    return self.value_head(x)
```

Step 3: Define the Proximal Policy Optimization (PPO) Agent

Define the PPO agent class, including the loss functions and update steps.

```
import numpy as np


class PPOAgent:

    def __init__(self, policy_network, value_network, optimizer, gamma=0.99, epsilon=0.2):

        self.policy_network = policy_network

        self.value_network = value_network

        self.optimizer = optimizer

        self.gamma = gamma

        self.epsilon = epsilon


    def compute_loss(self, states, actions, advantages, old_probs, returns):

        # Compute new probabilities and values

        probs = self.policy_network(states)

        values = self.value_network(states)


        # Compute advantages and value loss

        advantages = advantages[:, np.newaxis]

        value_loss = tf.reduce_mean(tf.square(returns - values))
```

115

```
# Compute policy loss

new_probs = tf.reduce_sum(actions * probs, axis=1,
keepdims=True)

old_probs = tf.clip_by_value(old_probs, 1e-10, 1.0)

new_probs = tf.clip_by_value(new_probs, 1e-10, 1.0)

ratio = new_probs / old_probs

surr1 = ratio * advantages

surr2 = tf.clip_by_value(ratio, 1 - self.epsilon, 1 +
self.epsilon) * advantages

policy_loss = -tf.reduce_mean(tf.minimum(surr1, surr2))


# Total loss

loss = policy_loss + 0.5 * value_loss


return loss


    def train_step(self, states, actions, advantages, old_probs,
returns):

        with tf.GradientTape() as tape:

            loss = self.compute_loss(states, actions, advantages,
old_probs, returns)


            gradients          =          tape.gradient(loss,
self.policy_network.trainable_variables          +
self.value_network.trainable_variables)

            self.optimizer.apply_gradients(zip(gradients,
self.policy_network.trainable_variables          +
self.value_network.trainable_variables))

        return loss
```

Step 4: Training Loop

Implement the training loop using PPO.

```python
import gym
import scipy.signal


# Create the CartPole environment
env = gym.make('CartPole-v1')


# Hyperparameters
num_episodes = 500
gamma = 0.99
epsilon = 0.2
learning_rate = 0.001


# Initialize the networks and optimizer
policy_network                              =
PolicyNetwork(num_actions=env.action_space.n)
value_network = ValueNetwork()
optimizer = tf.optimizers.Adam(learning_rate)
ppo_agent  =  PPOAgent(policy_network,  value_network,
optimizer, gamma, epsilon)


# Training loop
for episode in range(num_episodes):
    states, actions, rewards, old_probs = [], [], [], []
    state = env.reset()
    episode_reward = 0


    with tf.GradientTape() as tape:
```

```
for t in range(1, env.spec.max_episode_steps + 1):

    state = state.reshape(1, -1).astype(np.float32)

    prob = policy_network(state)

    action   =   np.random.choice(env.action_space.n,
p=prob.numpy()[0])


    states.append(state)

    actions.append(action)

    old_probs.append(prob.numpy()[0, action])


    state, reward, done, _ = env.step(action)

    rewards.append(reward)


    episode_reward += reward


    if done:

        break


states = np.vstack(states)

advantages   =   np.array(rewards)   +   gamma   *
np.append(0,
ppo_agent.value_network(states).numpy().flatten()[:-1])   -
ppo_agent.value_network(states).numpy().flatten()

returns = scipy.signal.lfilter([1], [1, -gamma],
rewards[::-1])[::-1]


states = np.vstack(states)

actions   =   tf.keras.utils.to_categorical(actions,
env.action_space.n).astype(np.float32)

old_probs = np.array(old_probs)
```

```
    loss = ppo_agent.train_step(states, actions, advantages,
old_probs, returns)


    print(f"Episode {episode + 1}/{num_episodes}, Total
Reward: {episode_reward}, Loss: {loss.numpy()}")


# Close the environment

env.close()
```

How to Run the Project:

Install the required libraries (**gym** and **tensorflow**).

Create policy and value networks.

Define the PPO agent.

Implement the training loop using PPO on the CartPole environment.

This project provides a basic example of implementing Proximal Policy Optimization (PPO) using TensorFlow. Customize the neural network architecture, hyperparameters, and other components based on your specific requirements and the characteristics of your chosen environment.

10.4 TensorFlow in RL Research

TensorFlow plays a crucial role in RL research, enabling the implementation and experimentation of cutting-edge algorithms. Researchers leverage TensorFlow's flexibility to create novel RL architectures, test hypotheses, and push the boundaries of what is possible in reinforcement learning.

10.5 TensorFlow Model Deployment in RL Environments

Deploying RL models in production environments involves considerations such as real-time interaction, integration with physical systems, and robustness. TensorFlow Serving, TensorFlow Lite, and other deployment tools can be adapted to serve RL models efficiently.

10.6 Summary

TensorFlow's integration with OpenAI Gym, the presence of TensorFlow Agents library, and the flexibility for implementing custom RL algorithms make it a powerful tool for Reinforcement Learning. Whether you are developing and experimenting with RL algorithms, conducting research, or deploying RL models in real-world applications, TensorFlow provides a comprehensive and adaptable framework for the entire RL pipeline. As you explore the field of Reinforcement Learning, TensorFlow will serve as a valuable companion in creating intelligent agents capable of learning from and interacting with their environments.

Chapter 11: Troubleshooting and Optimization in TensorFlow

Ensuring the optimal performance and robustness of machine learning models is crucial in production environments. TensorFlow provides various tools and techniques for troubleshooting issues and optimizing models. In this chapter, we'll explore common challenges, debugging strategies, and optimization techniques to enhance the efficiency of your TensorFlow-based projects.

11.1 Debugging TensorFlow Models

11.1.1 TensorBoard for Visualization

TensorBoard is a powerful tool for visualizing various aspects of your TensorFlow model during training. You can use it to monitor metrics, visualize the graph structure, and inspect the distribution of weights.

```
# Example: Logging metrics in TensorFlow for TensorBoard

import tensorflow as tf

from tensorflow.keras.models import Sequential

from tensorflow.keras.layers import Dense


model = Sequential([
```

```
Dense(128, activation='relu', input_shape=(784,), name='dense_1'),

    Dense(10, activation='softmax', name='dense_2')

])
```

```
# Compile the model

model.compile(optimizer='adam',

        loss='sparse_categorical_crossentropy',

        metrics=['accuracy'])
```

```
# Create a TensorBoard callback

tensorboard_callback    =    tf.keras.callbacks.TensorBoard(log_dir='./logs',
histogram_freq=1)
```

```
# Train the model with TensorBoard callback

model.fit(train_data, epochs=5, callbacks=[tensorboard_callback])
```

11.2 Model Optimization

11.2.1 Quantization

Quantization reduces the precision of weights and activations in a model, leading to smaller model sizes and faster inference. TensorFlow provides tools for quantizing models.

```
import tensorflow as tf

from tensorflow_model_optimization.python.core.quantization.keras import
quantize_model
```

```
# Example: Quantizing a Keras model

quantized_model = quantize_model(model)
```

11.2.2 Pruning

Model pruning involves removing less important weights, resulting in a sparser and more efficient model. TensorFlow Model Optimization provides pruning tools for Keras models.

```
import tensorflow as tf

from    tensorflow_model_optimization.python.core.sparsity.keras    import
pruning_schedule


# Example: Pruning a Keras model

pruning_params = {

    'pruning_schedule':                 pruning_schedule.ConstantSparsity(0.5,
begin_step=0, frequency=100)

}


pruned_model = tf.keras.models.clone_model(model)

pruned_model  =  tfmot.sparsity.keras.prune_low_magnitude(pruned_model,
**pruning_params)
```

11.3 Performance Optimization

11.3.1 GPU Acceleration

TensorFlow supports GPU acceleration, allowing you to speed up model training and inference. Ensure that TensorFlow is properly configured to use the available GPUs.

```
# Example: Configuring TensorFlow to use GPU

physical_devices = tf.config.list_physical_devices('GPU')

if physical_devices:

    tf.config.experimental.set_memory_growth(physical_devices[0], True)
```

11.3.2 Batch Size and Data Pipeline Optimization

Adjusting the batch size and optimizing the data pipeline can significantly impact training performance. Experiment with different batch sizes and use TensorFlow's **tf.data** API for efficient data loading.

Example: Configuring data pipeline for performance

batch_size = 64

train_data = tf.data.Dataset.from_tensor_slices((x_train, y_train)).shuffle(50000).batch(batch_size).prefetch(tf.data.experimental.AUTOTUNE)

11.3.3 Mixed Precision Training

Mixed precision training uses lower precision data types (e.g., float16) for some operations, reducing memory usage and speeding up training. TensorFlow provides tools for mixed precision training.

import tensorflow as tf

from tensorflow.keras import mixed_precision

Example: Mixed precision training

policy = mixed_precision.Policy('mixed_float16')

mixed_precision.set_global_policy(policy)

Create and compile the model

model = create_your_model()

model.compile(optimizer='adam', loss='sparse_categorical_crossentropy', metrics=['accuracy'])

11.4 Summary

Troubleshooting and optimizing TensorFlow models involve a combination of debugging techniques and performance optimization strategies. Utilizing tools like TensorBoard for visualization, incorporating assertions and debugging functions, and applying optimization techniques such as quantization, pruning, GPU acceleration, and mixed precision training contribute to the overall success of your machine learning projects. Regular monitoring, experimentation, and adaptation to the specific characteristics of your models and datasets are key practices in maintaining efficient and robust TensorFlow-based solutions.

Appendix: Useful TensorFlow Resources

Certainly! Here's a list of useful TensorFlow resources that can help you learn, explore, and master TensorFlow:

Online Documentation and Guides:

- TensorFlow Official Documentation: The official documentation is comprehensive and covers everything from basics to advanced topics.
- TensorFlow Tutorials: A collection of hands-on tutorials covering various aspects of TensorFlow, including image classification, text generation, and more.
- TensorFlow API Documentation: In-depth documentation for TensorFlow's APIs, providing details about functions, classes, and modules.

Books:

"Hands-On Machine Learning with Scikit-Learn and TensorFlow" by Aurélien Géron: A highly recommended book that covers both Scikit-Learn and TensorFlow for machine learning.

- **"TensorFlow for Deep Learning" by Bharath Ramsundar and Reza Bosagh Zadeh**: Focuses on deep learning using TensorFlow and covers practical applications.

Courses and Learning Platforms:

- TensorFlow in Practice Specialization on Coursera: A series of courses by deeplearning.ai covering TensorFlow for various applications.
- TensorFlow: Advanced Techniques Specialization on Coursera: A more advanced specialization by deeplearning.ai, delving into advanced TensorFlow concepts.
- TensorFlow 2 for Deep Learning Specialization on Coursera: A specialization focused on TensorFlow 2, covering various deep learning topics.

Blogs and Articles:

- TensorFlow Blog: Official TensorFlow blog with updates, tutorials, and case studies.
- Medium - TensorFlow: TensorFlow articles on Medium covering a range of topics.
- GitHub Repositories:
- TensorFlow GitHub Repository: The official GitHub repository for TensorFlow, where you can find source code, issues, and community contributions.
- TensorFlow Models Repository: A repository containing various pre-trained models, example implementations, and research models.